LOW-SLOPE ROOFING II

Low-Slope Roofing II

National Council of Architectural Registration Boards (NCARB)
1801 K Street, NW, Suite 1100-K
Washington, DC 20006
(202) 783-6500
web site: www.ncarb.org

ISBN 0-941575-38-1
Printed in the United States of America

This monograph was published in December 2001.

FOREWORD

Since November 1993, NCARB's Professional Development Program (PDP) has published a series of monographs addressing topics of particular interest to professionals in architecture and related fields. This monograph on low-slope roofing is the eleventh in this ongoing effort. This program is specifically designed to help architects meet the continuing education and professional development requirements of state boards for continued professional licensure. It is, however, also suitable for other professionals as well, either to meet the continuing education requirements of their particular profession or to broaden their general knowledge of the relevant subject matter. In addition, monographs published by NCARB are accepted by The American Institute of Architects (AIA) to meet its continuing education requirements for maintenance of membership in the Institute.

This monograph is the second of two dealing with the topic of weather protection on low-slope roof decks. The first monograph focused on built-up roof systems. This monograph focuses on other roof systems. Together, they provide basic information about the design of low-slope roof assemblies for practicing architects, students and others.

NCARB's PDP Committee selects the topics, reviews the content, and prepares the quiz questions for each monograph. Under the Committee's direction, NCARB has published monographs on such diverse topics as seismic mitigation, sustainable architecture, subsurface conditions and professional conduct. NCARB will shortly publish additional monographs entitled *Design within a Community Context* and *Post-Occupancy Evaluation.*

Along with developing new monographs, NCARB's PDP Committee reviews titles currently in print every six years for timeliness and applicability of general concepts. Where appropriate, the monographs are revised and republished.

NCARB welcomes your comments and questions about this monograph and about the monograph program in general. Please let us know of any other topics that you would like to see addressed. We thank you for your participation.

ABOUT THE AUTHOR

Thomas Lee Smith, AIA, RRC, is president of TLSmith Consulting Inc., in Rockton, Illinois (www.tl-smith.com). As a sole practitioner, he specializes in architectural technology and research, with an emphasis on roof systems. Smith is a licensed architect and holds the NCARB Certificate. He is also a registered roof consultant. His interest in roofing began in 1974. From 1988 to 1998, he was the research director for the National Roofing Contractors Association (NRCA). Prior to that time, he was in private practice in California and Alaska. He has designed roofs from the arctic to the tropics.

Smith is familiar with all of the low- and steep-slope roofing systems that are commonly used in the United States. He has a strong theoretical understanding and practical experience with many materials, systems and issues associated with roof performance. He is an internationally recognized expert on roofing technology. In particular, he is recognized for his expertise related to wind performance of roof systems.

In addition to the AIA and NRCA, Smith is a member of the American Association for Wind Engineering (AAWE), American Society for Testing and Materials (ASTM), The Construction Specifications Institute (CSI), Roof Consultants Institute (RCI) and the Southern Building Code Congress International (SBCCI).

Since 1990, Smith has served on the American Society of Civil Engineers (ASCE) committee that is responsible for *Minimum Design Loads for Buildings and Other Structures (ASCE 7),* and he is a member of its subcommittee on wind loads. He was elected to AAWE's board of directors in 1998. Since 1989, he has been on the joint committee on roofing materials and systems for the International Council for Research and Innovation in Building and Construction (CIB) and the International Union of Testing and Research Laboratories for Materials and Structures (RILEM). Beginning in 1988, he has been on the faculty of the Roofing Industry Educational Institute (RIEI). And from 1989 to 1998, he was a member of the Technology Assessment and Advisory Council for the Colorado State University/Texas Tech University Cooperative Program in Wind Engineering.

Smith has authored more than 100 magazine articles, numerous papers for national and international technical conferences, a chapter in two books, and various technical publications. In addition, he wrote *Roofing Standards, Considerations and Criteria,* which was published in 1986 for the State of Alaska.

INTRODUCTION

Not too many years ago, it was relatively easy for architects to adequately serve their clients' low-slope roofing needs. Up until the mid-to-late 1970s, almost all low-slope roofs were asphalt or coal tar built-up roofs *(figure 1)*. However, during the last two decades of the 20th century, other types of low-slope roof systems began to compete with traditional built-up roofs (BUR). BUR is still viable and will undoubtedly remain so for years to come. According to the NRCA's *1999-2000 Annual Market Survey,* BUR captured 19.5 percent of new construction and 21.3 percent of reroofing in the 1999 U.S. low-slope market. However, the challengers to BUR will also be with us for the foreseeable future.

FIGURE 1

Coal tar is poured and aggregate surfacing is shoveled onto a small built-up roof.

FIGURE 2

A sprayed polyurethane foam roof system is being applied with a robotic sprayer.

These challengers—modified bitumen, single-ply, sprayed polyurethane foam *(figure 2)*, metal, and liquid-applied systems—are unique. Only the modified bitumen systems are related to BUR; the other low-slope alternatives are radically different. To select the most appropriate roof system for a project, the architect should ideally have at least a general understanding of the available membrane options. This is a formidable task. An earlier monograph from NCARB, *Low-Slope Roofing I,* provided information on BUR. This companion document provides information on the other low-slope options.

SCOPE OF MONOGRAPH

Low-sloped roofs are defined as those roofs with a slope less than or equal to 3:12 (25 percent). However, with the exception of metal roofs, most low-slope roofs have a slope of about 1¼:12 (2 percent). The primary purpose of *Low-Slope Roofing II* is to present all the low-slope roofing alternatives to BUR and discuss their salient design considerations (most of which are also applicable to BUR). The monograph, however, also has seven chapters that are applicable to all low-slope roofing. These chapters include: "Reroofing Considerations," "Sustainable Design Considerations," "System Selection Criteria," "Warranty Considerations," "Key Elements of Specifications and Drawings," "Construction Contract Administration" and "Problems after Job Completion."

LOW-SLOPE ROOFING II

Low-Slope Roofing I discusses various types of roof decks and roof insulations. Both the deck and roof insulation strongly influence the roof assembly. Although the materials themselves are discussed in the first monograph, this monograph discusses the relationships between these components and the BUR alternatives. Study of both monographs is encouraged.

SIGNIFICANCE TO ARCHITECTS

To provide high-quality design service to clients, someone in the architect's office should possess a general knowledge of all low-slope roof system options and roof design criteria. These monographs can be useful in those offices where the knowledge level needs improvement. The knowledge acquired from them should benefit the firm's clients. And the knowledge may be instrumental in avoiding roof-related litigation and damage to the firm's reputation.

PROJECT DELIVERY

Delivering a successful roof to the building owner involves two distinct phases. The first phase is the design process. In this phase, the architect needs to have a basic understanding of roof membrane materials and system options *(see chapter entitled "Membrane Materials," page 7)*, and an understanding of roof design considerations *(see chapters entitled "Design Considerations," page 28, "Reroofing Considerations," page 66, and "Sustainable Design Considerations," page 85)*. After identifying the project's requirements, a roof system should be selected that optimally responds to the integration of the project's requirements and the system selection criteria discussed in the chapter by that name. After the roof system is selected, the specifics of the system (such as insulation type[s] and thickness, fastener patterns, and warranty requirements) are developed and details are designed (drawing upon information presented in the chapters "Membrane Materials" through "Warranty Considerations"). This phase is culminated with the preparation of specifications and drawings that communicate the architect's design concept and requirements to a professional roofing contractor for execution of the work *(see chapter entitled "Key Elements of Specifications and Drawings," page 109)*.

The second phase is construction contract administration, addressed in the chapter by that name. In addition to traditional activities, such as submittal review and field observation, the architect should also inform the building owner about the importance of semi-annual roof inspections and routine maintenance.

LIMITATIONS OF MONOGRAPH

The monographs are intended to give a relatively brief introduction to low-slope roofing. They are intended to address the basics: they do not delve deeply into the subject; otherwise, they would literally be a few thousand pages. After gaining a general understanding of the low-slope roof options and various issues associated with them, an architect has a choice to make: Either elect to further expand his or her skills and knowledge, or work with professional roofing contractors or roof consultants. Years ago, it was uncommon for architects to work with a roof consultant or call upon a trusted contractor for advice. But the complexities brought on by the BUR alternatives now demand the inclusion of a roof consultant as part of the design team if this expertise is not developed within the architect's office.

At the very least, architects should have some key reference materials in their office. These materials are discussed within the following chapters. If an architect desires to develop his or her expertise rather than use consultants, the firm's roofing library needs to expand beyond the key reference materials, and it needs to grow as the body of technical literature increases. The architect should also subscribe to at least one of the roofing industry's monthly periodicals *(see "Resources," page 165).* In addition to acquiring publications, it is also imperative for the architect to attend seminars. RIEI has a superb four-day in-depth basic course on roofing technology. It also has several one- and two-day courses on different types of roof systems, as well as topics such as quality assurance (field observation) and evaluating problem roofs.

In addition to having a good library (and reading what is in it) and attending quality seminars, it is also important to spend time on roofs with knowledgeable people who can provide direct hands-on teaching. By doing so, wisdom (that blend of knowledge and judgment that comes with experience) can be developed. The following chapters provide a big step on this life-long journey.

MEMBRANE MATERIALS

When specifying low-slope roof assemblies, architects have many deck, insulation and roof membrane materials from which to choose. The various assembly options can fulfill a wide range of functions, from energy conservation to wind, hail, fire and water resistance. Their ability to provide these functions throughout their expected service life depends on good design, quality materials, good application and a commitment by building owners to have their roofs maintained.

This chapter provides an overview of the membrane materials available for all low-slope roofing systems, except for built-up roofing (BUR), which is addressed in NCARB's *Low-Slope Roofing I.* Combining the membrane with the other roof assembly components is addressed later in the chapter entitled "Design Considerations." Roof system selection criteria are discussed in a subsequent chapter, "System Selection Criteria."

The term *roof assembly* includes the roof deck, vapor retarder (if present), roof insulation (if present) and the roof membrane. The term *roof system* refers to the vapor retarder (if present), roof insulation (if present) and the roof membrane.

Materials for the following types of systems are included in this chapter:

- modified bitumen
- single-ply
- sprayed polyurethane foam
- liquid-applied
- metal panels.

At the end of this chapter, several publications are listed that provide further information on these systems, such as historical development and manufacturing processes.

MODIFIED BITUMEN (MB)

MB membranes are typically composed of polymer-modified asphalt, although some polymer-modified coal tar products are also available (in North America, the term *bitumen* refers to both asphalt and coal tar). Polymers are added to bitumen to enhance various properties of the bitumen, such as flexibility, cold-weather ductility, elasticity and resistance to high-temperature flow.

MB membranes are related to BUR, and they commonly employ BUR components (such as base sheets). Prefabricated MB sheets are typically used to construct base flashings for BUR systems. MB membranes exhibit general toughness and resistance to abuse.

The quality of MB products is highly dependent on the quality and compatibility of the bitumen and polymers, and the recipe used during the blending process. Some MB sheets are of very high quality, while other sheets are considered commodity products that will require greater maintenance, offer a reduced service life or both.

The Midwest Roofing Contractors Association (MRCA) and NRCA published a research report on the field performance of a large number of modified bitumen membrane roofs located in various areas of the United States (MRCA 1996). Architects considering specifying MB membranes are encouraged to read the report.

There are four primary types of MB sheets, each identified by its particular polymer content. In addition, sheets can be manufactured with polyalphaolefin (PAO) polymer, but this type of sheet is not common in the United States. The four common types are described below.

Atactic Polypropylene (APP)

APP is blended with asphalt and fillers. The mixture is then factory-fabricated into rolls that are typically one meter wide. The prefabricated sheet, commonly referred to as a *cap sheet,* is typically reinforced with fiberglass, polyester or a combination of both. The sheets are available smooth (in other words, unsurfaced); embedded with mineral granules of a variety of colors; or factory-surfaced with metal foil such as aluminum, copper or stainless steel. The aluminum foil is available in colored finishes. The finished membrane also can be surfaced with aggregate, although this is uncommon. Unsurfaced sheets range in thickness from around 120 to 200 mils [3 to 5 mm].

To avoid surface cracking, the architect should specify a field-applied coating (such as aluminum-pigmented asphalt, asphalt emulsion or acrylic), factory-applied surfacing (granules or metal foil) or a sheet with protective reinforcement near the top.

APP MB membranes are typically composed of a base sheet and an APP cap sheet. The base sheet can be either an APP MB base sheet or an unmodified base sheet as used in BUR systems. Typically the base sheet is either fully adhered to the substrate with hot asphalt or cold adhesive, or it is torched or attached with mechanical fasteners. However, hot asphalt cannot be used to adhere an APP MB base sheet because the temperature of the mopping asphalt is insufficient to cause it to bond to the sheet as the addition of the APP to the asphalt in the APP MB sheet causes a substantial rise in the bitumen's softening point. If the base sheet is adhered in cold adhesive or torched, specify a base sheet that is manufactured for the applicable attachment method. For adhesive attachment, specify an adhesive formulated for adhering this type of sheet.

The APP cap sheet is either heat-welded (in other words, torched) to the base sheet *(figure 3),* or it is adhered in cold adhesive. Specify a cap sheet that is manufactured for the applicable attachment method. When the cap sheet is adhered in adhesive, the seams can either be heat-welded or joined by an adhesive. Some manufacturers recommend one particular method, and some allow both options. One advantage of heat-welding is that the seam is immediately watertight and not susceptible to being opened by wind or membrane thermal movements. However, mechanics need to exercise care when torching because of the solvent-bearing adhesive nearby.

FIGURE 3

This APP modified bitumen cap sheet is being applied with a multi-head torch trolley. The seam is being rolled with a weighted roller.

Sometimes one or more fiberglass ply sheets (as used in BUR) are mopped to the base sheet and then the cap sheet is installed. In this case, an APP MB base sheet should not be specified. This configuration is often referred to as a hybrid system, as it essentially is a BUR with a MB cap sheet. The additional ply sheets provide additional redundancy.

APP MB membranes can also be used in a protected membrane roof (PMR) configuration *(figure 4)*. In a PMR, extruded expanded polystyrene (XEPS) insulation specifically manufactured for this purpose is placed over the roof membrane. The insulation is protected from ultraviolet radiation (UV) and wind blow-off by concrete pavers or large aggregate. When aggregate is selected, the architect needs to specify a filter fabric between the aggregate and insulation in order to keep the aggregate from getting into the board joints and underneath the boards.

FIGURE 4
PMR Configuration

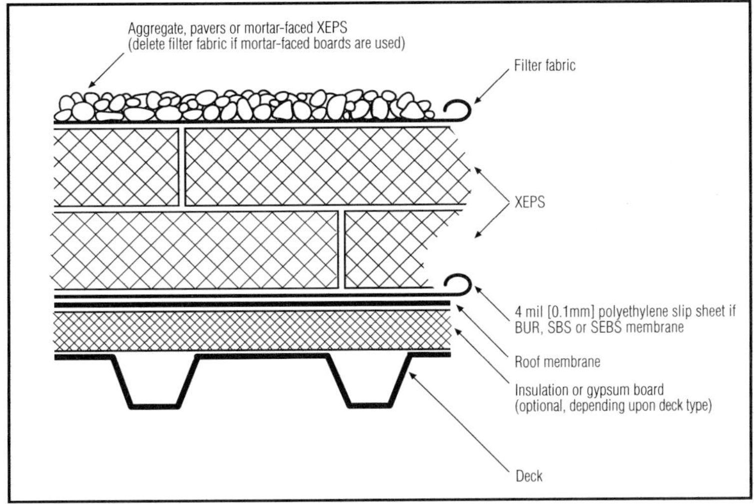

Alternatively, insulation boards with a factory-applied mortar surface may be specified *(figure 5)*. Aggregate-surfaced PMRs are limited to a maximum slope of 1½:12 (4 percent). Pavers and mortar-faced boards are limited to a maximum slope of 2:12 (17 percent).

FIGURE 5
This PMR system is composed of two layers of XEPS boards. The top board has a factory-applied mortar surface. Note that a strip of flexible foam, typically used to fill concrete expansion joints, is placed between the parapet base flashing and the XEPS to prevent the mortar from rubbing against the base flashing.

The following ASTM product standards are applicable to APP MB sheets:

- D 6222: Polyester-reinforced cap sheet, Type I and Type II (Type II can be considered a more robust sheet), with or without granules (Grade G or S).
- D 6223: Combination of polyester- and fiberglass-reinforced cap sheet, Type I and Type II (Type II can be considered a more robust sheet), with or without granules (Grade G or S).
- D 6509: APP MB base sheet, fiberglass-reinforced.

Styrene-butadiene-styrene (SBS)

SBS is blended with asphalt and fillers. The mixture is then factory-fabricated into rolls that are typically one meter wide. The prefabricated sheet is typically reinforced with fiberglass, polyester or a combination of both. The sheets are available smooth; embedded with mineral granules of a variety of colors; or factory-surfaced with metal foil such as aluminum, copper or stainless steel. The aluminum foil is available in colored finishes. The finished membrane can also be surfaced with aggregate, although this is uncommon. Unsurfaced sheets range in thickness from around 80 to 160 mils [2 to 4 mm].

SBS MB is susceptible to premature deterioration when exposed to UV radiation. A factory-surfacing, therefore, is typically specified for cap sheets. Otherwise, the architect should specify a field-applied coating and the architect should advise the building owner to have the membrane periodically recoated.

One manufacturer produces a modified coal tar product that is similar to SBS modified asphalt sheets although it uses a different polymer modifier. This product entered the marketplace in the late 1990s. There is no ASTM standard for this product.

At least one manufacturer produces liquid-applied SBS modified asphalt. This is described below with the other liquid-applied membranes.

SBS MB membranes are typically composed of a base sheet and an SBS cap sheet. The base sheet can be either an SBS MB base sheet or an unmodified base sheet as used in BUR systems. Typically the base sheet is either fully adhered to the substrate with hot asphalt or cold adhesive, or it is torched or attached with mechanical fasteners. If the base sheet is adhered in cold adhesive or torched *(figure 6)*, specify a base sheet that is manufactured for the applicable attachment method. For adhesive attachment, specify an adhesive formulated for adhering this type of sheet.

FIGURE 6
This SBS modified bitumen base sheet is being applied with a hand-held torch.

The SBS cap sheet is adhered to the base sheet in hot asphalt or cold adhesive, or it is heat-welded to the base sheet. Specify a cap sheet that is manufactured for the applicable attachment method. When the cap sheet is adhered in adhesive, the seams can either be heat-welded or joined by an adhesive. Some manufacturers recommend one particular method, and some allow both options. As with APP MB, heat-welding SBS MB creates a seam that is immediately watertight and not susceptible to being opened by wind or membrane thermal movements, but the solvent-bearing adhesive nearby means that mechanics need to exercise care when torching. Sometimes one or more fiberglass ply sheets (as used in BUR) are placed between the base sheet and the cap sheet. As previously mentioned, this configuration is often referred to as a hybrid system. The additional ply sheets provide additional redundancy.

SBS MB membranes can also be used in a PMR configuration. If a PMR system is specified, a minimum 4 mil [0.1 mm] polyethylene slip sheet should be placed between the membrane and the extruded polystyrene to prevent the insulation boards from bonding to the membrane. Otherwise, membrane tearing could occur when the insulation floats during a rainstorm.

The following ASTM product standards are applicable to SBS MB sheets:

- D 6162: Combination of polyester- and fiberglass-reinforced sheet, Type I and Type II (Type II can be considered a more robust sheet), with or without granules (Grade G or S).
- D 6163: Fiberglass-reinforced sheet, Type I and Type II (Type II can be considered a more robust sheet), with or without granules (Grade G or S).
- D 6164: Polyester-reinforced sheet, Type I and Type II (Type II can be considered a more robust sheet), with or without granules (Grade G or S).
- D 6298: Fiberglass-reinforced sheet, metal-surfaced.

Styrene-isoprene-styrene (SIS)

SIS is blended with asphalt and fillers. The mixture is then factory-fabricated into either 3-feet-wide or 1-meter-wide rolls. The top of the prefabricated sheet is available with embedded mineral granules or a factory-laminated UV-protective surfacing, such as aluminum foil. The bottom surface has a release paper to keep the sheet from bonding to itself while rolled.

A similar product is commonly used under steep-slope roof coverings to provide ice-dam protection. However, the steep-slope underlayments do not have a UV-protective surfacing. SIS MB roof membranes currently capture a very small share of the low-slope market, but this membrane type may experience some growth.

These sheets are self-adhering. During application, the release paper is removed and the sheet is pressed into place.

There is an ASTM standard for self-adhering MB underlayment, but there is no standard for self-adhering MB roof membrane products.

Styrene-ethylene-butylene-styrene (SEBS)

SEBS is blended with asphalt in a factory. The SEBS-modified asphalt is then reheated at the job site in tankers or kettles that are specifically designed to avoid damaging the material. The hot modified asphalt is applied in a manner that is virtually identical to BUR, using alternating layers of fiberglass or polyester ply sheets and the modified asphalt. The membrane is then typically surfaced with aggregate as used on BUR systems.

The SEBS-modified mopping asphalt is extremely expensive. Although SEBS MB membranes that are well designed and installed should be expected to provide very reliable and long service, the great expense of this type of membrane is typically difficult to justify.

SEBS-modified mopping asphalt is specified in ASTM product standard D 6152.

Modified mopping coal tar was introduced in the mid-1990s. Until substantial field data on the performance of this new product becomes available, however, architects should be cautious of specifying it. There is no ASTM standard for this product.

SINGLE-PLY

The single-ply family of roof membranes is composed of thermoplastic and thermoset products. Single-ply sheets are factory-fabricated and installed in a single thickness. The sheets typically are around 35 to 95 mils [0.89 to 2.41 mm] thick, depending upon the specific type of membrane. Single-plies were introduced in North America in the 1960s and 1970s. Many of the early-generation products did not offer the reliability or length of service life that was expected. As a result, several of the early products are no longer produced. Some early products, however, evolved through changes in their formulation or system design and are still available today. In addition, new types of single-plies have been developed over the years.

Single-ply membranes are relatively easy to install on steep or complex roof slopes. Except for the ballasted systems described below, they are also very light weight in comparison to BUR or MB membranes (*see "System Selection Criteria" chapter for a comparison of system weights, page 92*). However, unless used in a PMR configuration, they do not offer nearly as much toughness and resistance to abuse as do BUR and MB membranes.

Attachment Methods

Described below are the five primary methods for securing single-plies to the roof deck or other substrate:

Fully adhered

The membrane is adhered to the substrate in a continuous layer of adhesive (*figure 7*). The type of adhesive varies with the type of membrane and substrate.

FIGURE 7

Several mechanics are rolling this fully adhered single-ply membrane into place. In the foreground, the EPDM membrane was pulled back so that adhesive could be applied to the backside of the membrane. On the left, adhesive has been applied to the insulation.

Ballasted

The membrane is loose-laid over the substrate and then covered with ballast to resist wind uplift (*figure 8*). Ballast can be large aggregate (for example, 1½ or 2½ inches nominal diameter [38 or 64 mm], depending upon design wind speed); concrete pavers weighing 18 to 25 psf [89 to 122 kg/m^2]; or specially designed lightweight interlocking concrete pavers weighing approximately 10 psf [49 kg/m^2]. Ballasted systems are limited to a maximum slope of 2:12 (17 percent).

If crushed aggregate is specified, the architect should also specify a stone-protection mat between the membrane and aggregate. Otherwise, foot traffic

FIGURE 8
Aggregate ballast is applied over an EPDM membrane.

FIGURE 9
This EPDM membrane was punctured by a piece of aggregate (at the black upside-down "v" in the center of the photograph). The end of an ink pen shows the scale.

FIGURE 10
A piece of aggregate in the center of the photograph shattered into several sharp fragments after installation on a ballasted single-ply.

over the ballast could cause sharp fragments to puncture the membrane. And these punctures are extremely difficult to find *(figure 9)*.

Although not a common practice, it is prudent for architects to specify a stone-protection mat when smooth aggregate is used because some sharp fragments are often among the smooth aggregates and because aggregates sometimes fracture into very sharp pieces after they have been installed *(figure 10)*. It is also a conservative practice to specify a mat to protect against abrasion and puncture from fragments during paver installation. A somewhat thinner mat is normally sufficient for paver-ballasted jobs.

Mechanically attached

To avoid long membrane tears in the event that the membrane is torn, it is prudent for architects to specify only reinforced membranes for mechanically attached systems. Moderate winds can cause a torn non-reinforced membrane to lift and billow, which can lead to a very long tear propagation.

The membrane is attached with various types of fasteners through the roof insulation (if present) and into the roof deck. A variety of attachment systems have been introduced over the years, many of which subsequently proved to be unreliable. The most common approaches currently used are:

- Tab-attached: Fasteners and plates (typically 2 inches [50 mm] in diameter) are placed within the membrane seam at intervals of about 6 to 18 inches [150 to 460 mm] on center, depending upon design wind conditions and system resistance *(figure 11)*. The spacing between fastener rows varies from around 5 to 10 feet [1.5 to 3 m] on center, depending upon membrane type and design wind conditions and system resistance.

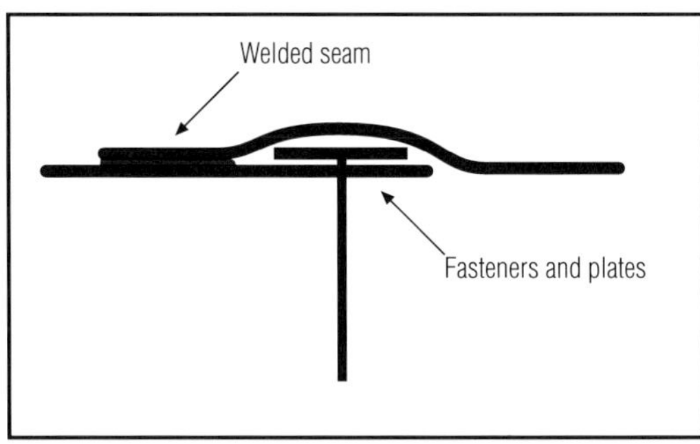

FIGURE 11
Tab-attached Membrane

- Batten-in-the-seam: Thin, flat metal or polymer battens are continuously placed within the seam and fastened to the deck. Alternatively, fasteners and plates are placed within the seam. The sheets on either side of the batten (or plates) are bonded together *(figure 12)*. This technique is typically used for mechanically attached EPDM systems because of their relatively low seam strength. This method is also known as a *balanced seam* or, in the case of thermoplastic membranes, *double-sided weld* because the sheets on both sides of the fasteners are bonded together.

  In another variation of this method, rows of fasteners and plates are placed on top of the membrane and fastened to the deck. The fasteners are then covered with a stripping ply of the membrane material.

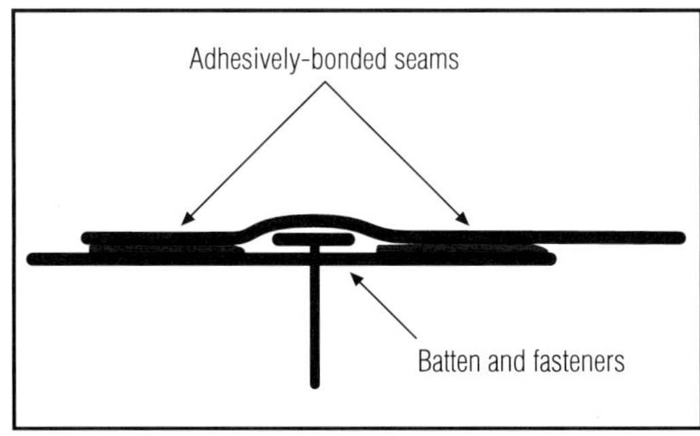

FIGURE 12
Batten-in-the-Seam Membrane Attachment
Fasteners and plates may be used in lieu of batten (check with manufacturer).

FIGURE 13
A bar-over attachment method is being used to mechanically attach this single-ply. After placing and seaming the field sheets, the crew placed bars over the PVC membrane and screwed them to the deck. A mechanic is cleaning the membrane on either side of the bar before welding on a stripping ply.

- Bar-over: Metal bars are placed at intervals on top of the membrane and then fastened to the deck *(figure 13)*. The bars are then covered with a stripping ply of the membrane material *(figure 14)*. These bars are thicker and more robust than the battens used for batten-in-the-seam attachment. Some bars are somewhat channel-shaped in order to provide enhanced stiffness. Field research after strong winds has revealed that, depending upon the bar design, row spacing and fastener spacing, this attachment method can be very reliable.

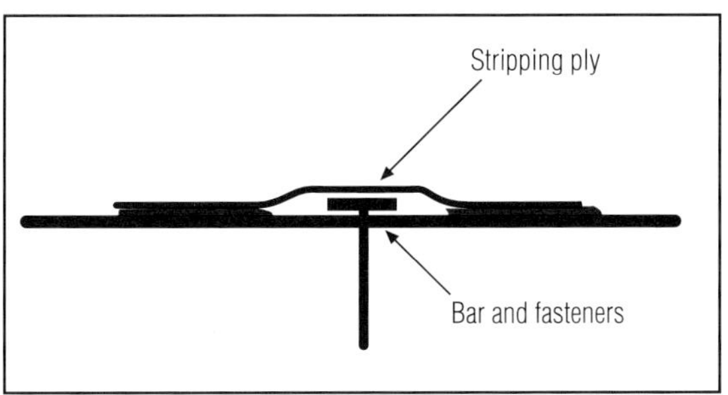

FIGURE 14
Bar-over Membrane Attachment

Protected membrane roof
This method is described above in the section on Modified Bitumen.

Loose-laid air-pressure equalization system
The membrane is fully adhered around the roof perimeter, but elsewhere the membrane is only loose-laid. This system should only be used over an air-impermeable roof deck or over an air retarder (*see chapter on "Design Considerations," page 28*). To compensate for minor air leakage between the membrane and the deck/air retarder, air-pressure equalization valves are installed at prescribed intervals. The valves are one-way: they allow air underneath the membrane to vent out, but outside air is prevented from flowing through the valve and underneath the membrane. When properly designed and installed, this type of attachment system can offer high-wind resistance—provided that holes are not subsequently cut through the deck/air retarder and left unsealed.

As with mechanically attached systems, it is prudent for architects to only specify reinforced membranes for air-pressure equalized systems.

Thermoplastic Products

Thermoplastic materials do not *cross-link,* or cure, during manufacturing or during their service life. It is therefore possible to weld these materials together. Typically field-fabricated seams are welded with robotic hot-air welders. Hand-held hot-air welders are used to weld seams at flashings and penetrations. Some manufacturers require the exposed edge of the seam to be sealed. Because the seams are welded, thermoplastic membrane seams are typically extremely reliable, resulting in a very low incidence of seam failures. Thermoplastic sheets are normally around 5 to 10 feet wide [1.5 to 3 m]. Some manufacturers weld the sheets together in the factory to form large sheets that are then welded together on the roof.

The types of membranes in the thermoplastic category are discussed below.

Polyvinyl chloride (PVC)
PVC membranes are among the oldest single-plies still available. Several early-generation sheets were unreinforced, and many of these roofs ended in premature catastrophic failure. Other early sheets were poorly formulated, too thin to provide a long service life, or both. However, the ASTM standard for PVC sheets was substantially strengthened in the mid-1990s. PVC membranes that comply with ASTM D 4434 Type II or Type III should be expected to provide reasonable service when properly designed, installed and maintained. As part of the D 4434 revisions, a Type IV product was added to the standard. While the Type II and III products are required to have a minimum thickness of 45 mils [1.1 mm], Type IV has a minimum requirement of 36 mils [0.9 mm]. Most of

the PVC membranes that are now produced have a thickness greater than 45 mils [1.1 mm], which provides a greater reservoir of plasticizer. This should result in longer service life.

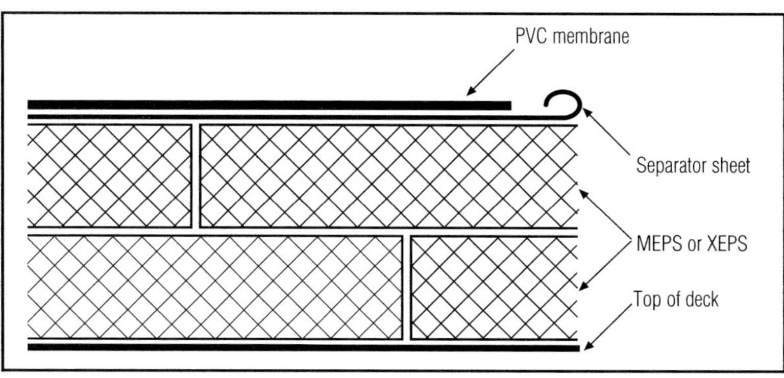

FIGURE 15
PVC Over Polystylene Insulation
Separator sheet is needed to avoid migration of plasticizers from PVC membrane.

If PVC membranes are in contact with polystyrene insulation, the polystyrene will cause the plasticizers in the membrane to leach out. To avoid such membrane embrittlement, a separator sheet needs to be installed between the membrane and the polystyrene *(figure 15)*. To avoid membrane damage, a separator is also needed to isolate PVC from asphalt and coal tar products.

Although several manufacturers offer ballasted PVC membranes, it is prudent for architects to not specify this attachment method. Fine dust particles from the ballast or particulate fall-out from the atmosphere commonly occur over the membrane within a few years after roof application. If these particles are clay-like, they too can leach plasticizers from the membrane.

PVC membranes are available in a wide variety of colors. This membrane is often selected for steep-slope roofs where a strong or unique color is desired.

PVC Alloys or Compounded Thermoplastics (also known as PVC blends)

These membranes are related to PVC membranes. They are primarily compounded from PVC, but they have additional polymers that provide somewhat different physical properties. Only a very small number of manufacturers make these products. There are no ASTM standards for these products. The primary types of membranes in this category are:

- Copolymer alloy (CPA): This membrane is reportedly compatible with oils and greases, animal fats, asphalt and coal tar.
- Ethylene interpolymer (EIP).
- Nitrile alloy (NBP).

Thermoplastic polyolefin (TPO)

TPO is the latest thermoplastic membrane introduced into the marketplace. It was commercialized in North America in the early 1990s, although it was introduced a few years earlier in Europe. It is formulated from polypropylene, polyethylene or other olefinic materials. Unlike PVC and PVC blends, it does not rely upon plasticizers for flexibility, so embrittlement due to plasticizer loss is of no concern, and it does not contain chlorine. TPO membranes are typically white.

There were seam problems with some of the early generation products. However, it appears that at least some of the products currently produced can be expected to deliver reasonable service life. In the late 1990s, TPOs experienced rapid growth in market share and several major manufacturers offered this product.

Field research studies on the performance of TPO membranes are limited. In the next few years, architects are urged to exercise caution if TPOs are specified. Products from manufacturers that have had experience with TPO should be expected to perform well. However, manufacturers that have recently introduced the product may experience start-up problems. An ASTM product standard is now being developed.

Ketone ethylene ester (KEE)

This membrane is also referred to as a tripolymer alloy (TPA), and the polymer is known by the trade name of Elvaloy. An ASTM product standard is now being developed.

Polyisobutylene (PIB)

Unlike the other thermoplastic membranes, PIB membranes are seamed with a tape and therefore are not as reliable as welded seams. PIB sheets are specified in ASTM product standard D 5019, Type II. (D 5019 Type I is for chlorosulfonated polyethylene, which is a thermoset product discussed later in this chapter.)

Chlorinated polyethylene (CPE)

This type of thermoplastic membrane is no longer available in North America.

Thermoset Products

Thermoset materials normally cross-link during manufacturing. Once cured, these materials can only be bonded together (for example, at a seam) with a bonding adhesive.

The types of membranes in the thermoset category are discussed on the next page.

Ethylene propylene diene monomer or terpolymer (EPDM)

EPDM is a synthetic rubber sheet. As of 2001, EPDM enjoys the largest market share of the single-plies in service in North America. EPDM membranes are extremely resistant to weathering, and they have very good low-temperature flexibility. However, EPDM is susceptible to swelling when exposed to aromatic, halogenated and aliphatic solvents, and animal and vegetable oils such as those exhausted from kitchens *(figure 16)*. On portions of roofs where the membrane may be exposed to these materials, an epichlorohydrin membrane can be specified over the EPDM as discussed below. EPDM membranes are suitable for airport buildings, provided liquid fuel is not spilled on the membrane.

The sheets are typically available in widths of 10, 20 and 45 or 50 feet [3, 6 and 14 or 15 m], and lengths up to 200 feet [61 m]. Hence, on large roofs with very few penetrations, this type of membrane can be very economical to install. Sheet thickness is typically 45 or 60 mils [1.1 to 1.5 mm]. Most EPDM sheets are black, although white sheets are available. The white sheets, however, are not nearly as resistant to weathering as the black sheets.

FIGURE 16

This mechanically attached EPDM swelled after being exposed to fuel oil.

EPDM is typically non-reinforced. As previously mentioned, however, it is prudent for architects to only specify reinforced sheets for mechanically attached and loose-laid air-pressure equalized applications. Reinforced sheets also offer some increased resistance to puncture and tearing when used in fully adhered and ballasted applications.

Another advantage to reinforced sheets is that they are not prone to shrinkage-induced problems. In ballasted, mechanically attached and air-pressure equalized systems that use non-reinforced sheets, the membrane will shrink over time. Although the shrinkage forces are relatively small, if the base securement, or *tie-in*, at roof perimeters and penetrations is not sufficient, the membrane can rupture *(figures 17 through 19)*. For further information on avoiding shrinkage-related problems, see "Shrinkage of EPDM Roof Membranes: Phenomenon, Causes, Prevention and Remediation" by Ralph Paroli and others.

In fully adhered applications, typically a contact adhesive is applied to the substrate and the sheet. After the adhesive dries, the sheet is mated with the substrate. A newer and faster method of application uses fleece-backed EPDM,

FIGURE 17
EPDM shrinkage caused the base securement detail to fail along the parapet. The base flashing now bridges from the top of the parapet out a few feet into the field of the roof. In this condition the membrane is very vulnerable to rupture.

FIGURE 18
This is similar to *figure 17* except that, on this roof, the top of the membrane has pulled out from underneath the metal counterflashing. The roof is no longer watertight in this area.

FIGURE 19
EPDM shrinkage caused the neoprene flashing along the metal edge flashing to rupture. The underlying field sheet tore around the nails at the flashing's horizontal flange.

which is set in a low-rise sprayed polyurethane foam adhesive.

Field seams are fabricated with either a liquid-applied adhesive or specially formulated tape. Early-generation tapes proved to be ineffective. Extensive research, however, has demonstrated the viability of tape-bonding. This is clearly the trend: some manufacturers now only offer tape-bonding. Although tapes offer performance advantages over liquid-applied adhesives, the contractor still needs to exercise care in cleaning the EPDM prior to tape application, priming the EPDM and diligently executing the seam work as recommended by the manufacturer.

EPDM sheets are specified in ASTM product standard D 4637. Type I sheets are non-reinforced, Type II sheets are reinforced and Type III sheets are fabric-backed.

Chlorosulfonated polyethylene (CSPE)

This membrane is also commonly known by the trade name Hypalon. Unlike the other thermoset membranes, CSPE is not cross-linked during manufacturing. During application, therefore, the seams are welded rather than adhered. The membrane begins to cure upon exposure and is soon cross-linked. Future repairs are then accomplished with adhesives. CSPE membranes are typically white. Much of the CSPE market share has been taken over by TPO membranes.

CSPE sheets are specified in ASTM product standard D 5019, Type I (D 5019 Type II is for PIB, a thermoplastic product). The ASTM standard uses the acronym CSM for CSPE.

Epichlorohydrin (ECH)

This sheet is similar in appearance to EPDM. ECH, however, is resistant to hydrocarbons, solvents, and many greases and oils, so it can be used in areas of the roof that are exposed to chemical discharges that are harmful to EPDM.

Because of its permeability, the ECH manufacturer recommends placing ECH over an EPDM membrane. Because it is so specialized, ECH is seldom used. Only one manufacturer produces it in North America. There is no ASTM standard for this product.

Neoprene (CR)

Neoprene is a trade name, but the name became so commonly used that it is now thought of as a generic term. The true generic term is polychloroprene. This membrane was the forerunner to EPDM. EPDM was not available with Class A fire resistance in its early years. Since neoprene sheets had a Class A rating, they were often specified in lieu of EPDM where greater fire resistance was desired. Neoprene was also used to flash early EPDM roofs. EPDM sheets eventually were formulated so that they too obtained a Class A rating, and EPDM flashing materials were developed. With these improvements to EPDM membranes, neoprene roof membranes are no longer produced in North America.

SPRAYED POLYURETHANE FOAM (SPF)

SPF, which dates back to the 1960s, is a very unique type of roof system. SPFs are constructed by spraying a two-part liquid onto a substrate. The two components mix after leaving the spray nozzle. The mixture expands and solidifies to form closed-cell polyurethane foam. The substrate can be either the roof deck, an existing roof membrane (provided the existing roof is suitable for re-covering), gypsum board or rigid insulation. The foam is applied with hand-held sprayers or by robotic sprayers *(figure 2 on page 3)*. Each pass (or lift) of foam is typically between ½ to 1½ inches [13 to 38 mm] thick. If a greater total thickness is desired, two or more passes are normally required. A minimum total thickness of 1 to 1½ inches [25 to 38 mm], depending upon the smoothness of the substrate, should normally be specified by the architect. The total thickness of the foam can be easily varied to provide slope for drainage.

Surfacings

The foam needs to be protected from UV radiation. This is typically accomplished by using one of the following surfacings:

Acrylic coating

This is the least expensive of the coatings and generally offers the shortest service life (although the best acrylics can last longer than some of the polyurethane coatings). With acrylics, re-coating is required about every 10 to 15 years, depending upon the quality of the coating material, application and climate. They are typically white. Acrylic roof coatings are specified in ASTM product standard D 6083.

Polyurethane coating

When properly formulated, this coating offers long service life. This can be the toughest coating available in terms of impact and tear resistance, although a wide range of physical properties is available in this product category. Both one-

and two-part coatings are available. One-part coatings are typically gray, although white is available. Two-part coatings are typically white. An ASTM product standard for one-part polyurethane roof coatings is now being developed. Upon completion of work on that standard, work is expected to begin on a standard for two-part polyurethane coatings.

Silicone coating

Silicone coatings offer exceptionally good weather resistance and long service life. These coatings are typically offered in a gray color, as silicone coatings pick up dirt (if a white silicone is installed, it will soon become gray). More than other coatings, silicone coatings are prone to being pecked by birds. To avoid the pecking, granules are commonly broadcast into the coating while it is wet.

FIGURE 20

Granules are being broadcast into wet coating on this SPF roof.

Mineral granules

Mineral granules (similar to those used to surface asphalt shingles) can increase the durability of a coating and provide greater slip-resistance to persons on the roof. Course sand can also be used for these purposes. Granules or sand are broadcast into a coating while it is wet *(figure 20).*

Aggregate surfacing

An alternative to coating the foam, because quality foam is quite resistant to liquid water, is to simply apply over the foam surface a layer of aggregate of the size used on BUR systems. In this scenario, at parapets and equipment curbs, one of the previously described coatings is applied on the vertical surfaces and out several inches onto the field of the roof.

Because water vapor can migrate through the foam, the aggregate surfacing option should not be specified in situations where the annual net vapor flow is downwards.

Other Considerations

In the early years of SPF development, the required physical properties of the foam and coatings were not established, application equipment was primitive, and design principles and application parameters were not well understood. As a result, many of the early jobs did not perform well. For the past several years, however, the tools have been available to allow architects and contractors to confidently specify and install these systems. One precautionary note: the worker who is performing the spraying must be very skilled and knowledgeable. If the qualifications of the contractor and the spray mechanic cannot be reasonably assured in a publicly bid situation, it is prudent for the architect to consider an alternative system.

SPF systems have several important attributes. They are exceptionally thermally efficient, since they do not have mechanical fasteners or insulation board joints, which create thermal bridges. Field research has demonstrated that they have exceptionally good wind resistance (Smith 1993). And, perhaps most important, the roof is not in imminent danger of leaking if the coating is weathered away *(figures 21 and 22)* or ruptured or the aggregate surface is displaced, provided that the penetration does not extend all of the way through the foam (which is generally unlikely). This attribute, exhibited by liquid-applied membranes as well, is in stark contrast with the other low-slope system options, in which leakage typically occurs if the membrane is punctured.

Architects should become familiar with this type of roof system, as it is a viable option for many roofing projects. It readily lends itself to complex roof shapes *(figure 23)*. In addition to the sources of information listed at the end of this chapter, *A Field and Laboratory Assessment of Sprayed Polyurethane Foam-Based Roof Systems* by Rene Dupuis and "Sprayed Polyurethane Foam Roofing: Overcoming Aesthetic Barriers" by Thomas Smith and Rene Dupuis are recommended for further reading, along with several publica-

FIGURE 21

The coating on this SPF roof had weathered away in some areas *(see figure 22)*. An electrical capacitance meter, however, did not detect moisture.

FIGURE 22

Where the coating weathered away on this SPF roof, about 1/8 inch [3 mm] of foam had also weathered away.

FIGURE 23

SPF roof system on a complex roof shape.

FIGURE 24

A liquid-applied membrane is being installed over plywood. A mechanic is applying flashing compound and reinforcing fabric over the panel joints.

tions by the Spray Polyurethane Foam Alliance (*see "Resources," page 165, for the SPFA web site, which lists their publications*).

Sprayed polyurethane roofing foam is covered by ASTM Standard C 1029, Type III and IV.

LIQUID-APPLIED

Liquid-applied membranes are applied directly to the structural deck (typically either concrete or plywood). Liquid-applied roof membranes are common in Guam, Puerto Rico and the U.S. Virgin Islands, but not in most parts of North America. The membrane is typically not reinforced, unless it is applied over plywood, in which case the joints are normally treated with a reinforcing fabric *(figure 24)*.

Field research has demonstrated that liquid-applied membranes have exceptionally good wind resistance. Also, if the membrane is punctured or torn, the water typically is unable to penetrate the deck (unless the puncture occurs over a crack or joint in the deck).

Buildings, particularly high-rises, located in areas with extremely high wind speeds are good candidates for liquid-applied membranes because of their reliable wind performance. If the membrane is punctured or torn by wind-blown debris, or if the membrane does de-bond (which is unlikely), the roof membrane material does not present as great a threat to other property or persons as do most other types of roof coverings. Care needs to be exercised in selecting parapet height, paver size and paver weight if a liquid-applied system is used in a PMR configuration in an area subjected to high design-wind conditions (*see the Wind Performance section in the "Design Considerations" chapter, page 36*).

Common liquid-applied materials are described below.

Polyurethane

This product is very similar to the polyurethane that is used to coat SPF. Different grades of roof membrane material are available, depending upon the severity of anticipated foot traffic.

Acrylic

This product is somewhat similar to the acrylic that is used to coat SPF. A very good slope (¼:12, or 2 percent, minimum in all areas) is required to avoid *birdbaths,* small pools of standing water that can prematurely degrade the membrane.

Hot-applied Rubberized Asphalt

This system uses hot asphalt that reportedly has been modified with SBS. It is applied at a somewhat lower temperature than that used for standard roofing asphalt and for SEBS modified mopping asphalt. The installed thickness is about 180 mils [4.6 mm]. A reinforcing fabric can be specified. To provide UV protection, a granule-surfaced modified bitumen cap sheet can be installed over the liquid-applied membrane while it is still hot. More commonly, however, this membrane is used in a PMR configuration. The PMR system is often used on high-rise roofs. This product is specified in Canadian General Standards Board (CGSB) standard 37.50-M89.

Silicone

The same material that is used to coat SPF can be used as a liquid-applied membrane.

METAL PANELS

Metal panels are not typically thought of as options for low-slope roofs. Some metal panel systems, however, can be used on very low slopes. Some manufacturers tout their systems as being suitable for slopes as low as ¼:12 (2 percent). NRCA, however, recommends a minimum slope of ½:12 (4 percent). The greater the slope, the more reliable the leakage protection. This section only addresses metal panels on slopes of 3:12 (25 percent) and less. Many more metal panel options are available for slopes greater than 3:12 (25 percent), but they are beyond the scope of this monograph.

When installed on low slopes, particularly slopes approaching ½:12 (4 percent) or less, a metal panel system needs to provide water infiltration protection all across the roof surface. Thus, low-slope metal panel systems should be designed and installed with the intent of making them membrane-like. To achieve this, the panel joints must be soldered or sealed together with sealant tape or sealant, or both. And fasteners that penetrate the panel at end-joint splices or flashings must be sealed with gasketed washers. It is more difficult to achieve a reliable and long-lasting watertight system on a very low-sloped roof with metal than it is with the other low-slope membrane materials.

Steel or aluminum panels are typically specified for low-slope standing seam panels. Copper is also available, but not commonly used since low-slope roofs are not normally visible. For corrosion protection on steel panels, it is prudent for architects to specify aluminum-zinc alloy (commonly known by the trade name Galvalume). Until the late 1990s, unpainted aluminum-zinc alloy panels had a factory-applied lubricant to facilitate roll-forming. The lubricant eventually weathers away, but installation smudges and fingerprints result in uneven appearance for awhile. A thin clear acrylic coat can be specified to provide a more even appearance and show the effects of weathering more gradually, as the acrylic weathers away. Acrylic-coated Galvalume is sold under trade names such as Galvalume Plus and Acrylume.

Normally low-slope panels are not painted. However, if the roof can be viewed from above, and it is desired to have a painted finish applied to either steel or aluminum panels, there are several finish options. A polyvinylidene fluoride (PVDF) finish, commonly known by the trade names of Kynar and Hylar, is typically specified since if offers a large range of colors and is extremely resistant to color change over time.

Avoid internal gutters and parapets at the eaves of low-slope metal roofs. Rather, it is less problematic to have the water flow over the end of the panels and fall directly to grade or drop into an external gutter that is below the plane of the panels. (*See the chapter on "Design Considerations" regarding icing, page 41.*)

The two primary approaches to low-slope metal roofs are described below.

Standing-Seam Hydrostatic or Water Barrier Systems

These panel systems are designed to resist water infiltration under hydrostatic pressure. The panels have standing seams, which raise the joint between the panels above the water line. The seam is sealed with sealant tape or sealant in case it becomes inundated with water backed up by an ice dam or driven by high wind.

Most hydrostatic systems are structural systems (in other words, the roof panel has sufficient strength to span between purlins or nailers). A hydrostatic architectural panel (which cannot span between supports) may be specified, however, if a solid deck is provided. Most architectural panels are *hydrokinetic,* or water-shedding, and therefore require a slope greater than 3:12 (25 percent).

Some panels have snap-together seams, while others are mechanically seamed with an electrically powered mechanical seaming tool. On slopes of ½:12 (4 percent) or less, it is prudent for the architect to specify mechanically seamed panels *(figure 25).* Some structural panels use a batten seam, but this type of seam is more commonly used on architectural panels.

There are two basic types of standing-seam panel profiles: the trapezoidal rib *(figure 26)* and the vertical rib *(figure 27).* Because of its appearance, the trapezoidal rib panel is typically used on industrial buildings and warehouses. It is difficult to make a trapezoidal panel watertight at hips and valleys.

In addition to the standing seam panels, *through-fastened panels* (also referred to as *box-rib panels*) with exposed fasteners *(figure 28)* are available for low-slope systems that are well sloped (for example, in excess of 2:12, or 17 percent). They should be considered hydrokinetic systems. This is a relatively inexpensive system. It has largely been replaced by standing seam systems, which eliminate leakage problems that are often associated with exposed fastener systems.

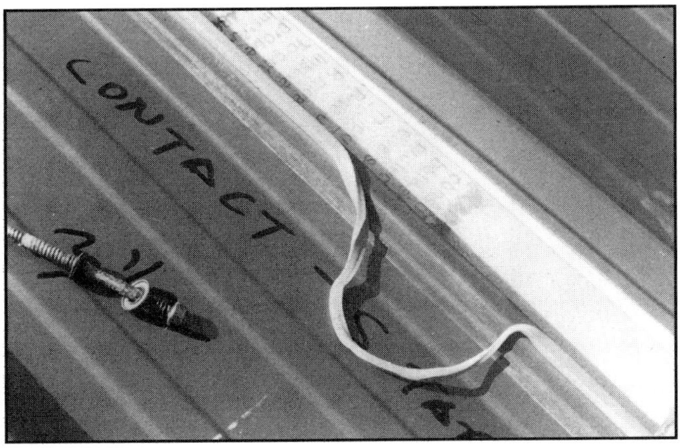

FIGURE 25

This standing seam panel system snaps together. A sealant tape was placed in the seam, but the top panel did not make contact with the tape. The seam was therefore susceptible to water intrusion. Panel systems that are seamed with an electrically powered mechanical seamer are less prone to this type of problem.

FIGURE 26

Trapezoidal Rib Standing Seam Panel

Ribs are typically 24 inches [610 mm] on center and 3 inches [75 mm] high.

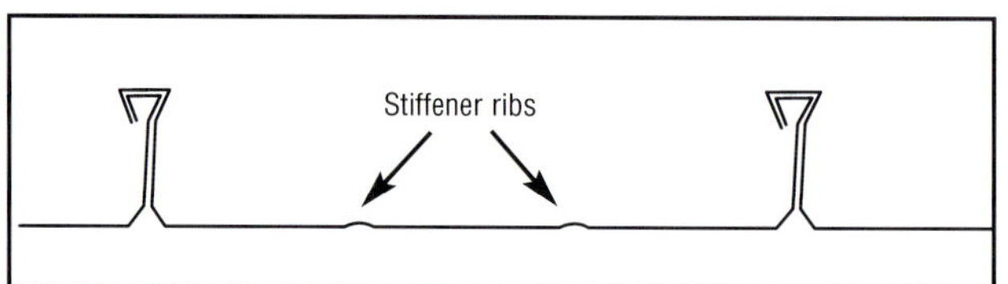

FIGURE 27

Vertical Rib Standing Seam Panel

Rib height and spacing vary with manufacturer.

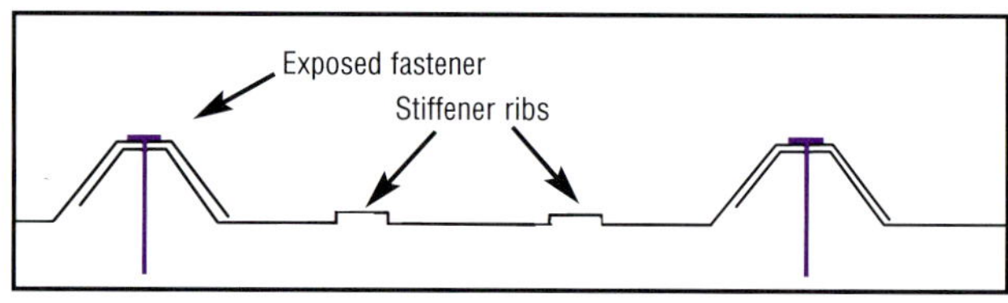

FIGURE 28

Through-fastened Panels

Primary ribs are typically 12 inches [305 mm] on center and 1½ inches [38 mm] high.

To secure the panels to the roof structure, it is prudent for the architect to specify a concealed clip system in lieu of an exposed fastener system because exposed fastener systems are more prone to leakage. In some clip designs, the clip lies over the first panel and the second panel is then placed over the clip and seamed. In such designs, at slopes of 3:12 (25 percent) or less, it is prudent for the architect to specify a bead of sealant under the clip before it is installed. Sealant tape is then installed over the panel rib and clip (or the overlying panel can have sealant pre-applied in the rib cavity). With this approach, there are no metal-to-metal joints. If sealant is not applied under the clip, leakage could occur at each clip location.

To avoid leakage problems at panel end-joint splices, it is preferable for the panels to be continuous from eave to ridge. This may necessitate job-site roll-forming. In some cases, full-length panels are impractical.

Structural standing-seam steel roof panel systems are specified in ASTM product standard E 1514. Structural standing-seam aluminum roof panel systems are specified in E 1637.

Flat Seamed Architectural Panels

This is also a hydrostatic, or water barrier, system. This traditional system requires a solid substrate. It also requires the use of metals that can be soldered, such as copper. This type of system is labor-intensive. Hence, it is relatively expensive. Because it demands diligent workmanship to provide long-term water protection, it is prudent for architects to not specify this system unless it is done so for architectural restoration purposes.

SOURCES OF ADDITIONAL INFORMATION

The following manuals and books have further information on low-slope membrane materials:

- *The NRCA Roofing and Waterproofing Manual:* This very comprehensive document has information on all of the membranes discussed in this chapter, with the exception of liquid-applied roofing membranes. The manual is updated approximately every seven years. A copy of the current manual should be in the office of every architect who designs low-slope roofs.
- *Low-Slope Roofing Materials Guide* by NRCA: This document, which is updated every other year, provides information on all of the membranes discussed in this chapter, with the exception of liquid-applied membranes. In addition, it includes information on insulation and fasteners. The guide lists manufacturers of the various types of products and provides information about the products. It provides a quick and convenient way to compare products within a membrane category.
- *Metal Roofing Systems Design Manual* by MBMA: This manual addresses metal roof systems.
- *The Manual of Low-Slope Roof Systems* by C.W. Griffin and R.L. Fricklas: This book provides information on all of the membranes discussed in this chapter.
- *The Science and Technology of Traditional and Modern Roofing Systems* by H.O. Laaly: This book provides information on all of the membranes discussed in this chapter.
- *Modified Bitumen Design Guide for Building Owners* by ARMA: This document only addresses modified bitumen systems.
- *Flexible Membrane Roofing: A Professional's Guide to Specifications* by SPRI: This document only addresses modified bitumen and single-ply systems.

DESIGN CONSIDERATIONS

Good long-term roof system performance depends on good design, materials, application, maintenance and repair. A significant shortcoming with any of these five elements can jeopardize performance of the roof system. Design, however, is the key element to achieving good long-term performance. Design inadequacies frequently cannot be compensated for by the other four elements. Good design, however, can compensate for other inadequacies to some extent.

The design of a roof system should consider the site's climate both during and after construction, the building's interior conditions (such as humidity, value of contents and operations) and characteristics of the building itself (such as the roof structure, size and shape). Developing a design that sufficiently accommodates the numerous demands placed on the roof system takes considerable effort. Several of the design considerations discussed in this chapter influence both the selection of the roof system (*see "System Selection Criteria" chapter, page 92*) and how the system is developed and detailed. The considerations that influence system selection vary from job to job, depending upon project location and requirements. Other design considerations are only applied after the system is selected. Reroofing and sustainable design considerations are discussed in subsequent chapters.

While other considerations may be applicable to some projects, the following design considerations include those that typically determine design success or failure:

- aesthetics
- geographical influences
- building codes
- Factory Mutual
- fire performance
- hail resistance
- wind performance
- roof drainage
- deck influences
- energy efficiency
- insulation influences
- vapor retarders
- self-drying roofs
- flashings and penetrations
- rooftop traffic
- building size and shape
- chemical resistance and compatibility
- insect attack
- construction above or adjacent to a roof
- green roofs
- steep-slope roofs
- maintenance
- essential facilities
- durability

- reliability
- roof system weight.

Other important items that architects need to consider include track record of the specified materials, cost and other non-technical factors as discussed in the "System Selection Criteria" chapter.

AESTHETICS

Low-slope roofs are usually not visually prominent, hence aesthetics is typically not an issue. However, low-slope roofs are visible from higher buildings *(figure 29),* and it is common to be able see the roof up close in buildings with multiple roof levels. Although the importance of aesthetics ranks low in the list of design considerations, aesthetics should not be forgotten about. In those cases where the roof can be viewed, consideration should be given to roof surface color, texture and reflectivity, in addition to the placement and screening of rooftop equipment.

FIGURE 29

This convention center roof can be seen from an adjacent hotel tower. The ballasted roof has bands of light- and dark-colored aggregate.

Depending upon the system selected, color can be provided by mineral granules, aggregate, pavers, field-applied coatings, factory-applied paint (in the case of metal panels) or the inherent color of the roof surface material itself. The available color palette is quite extensive. Vegetation can also provide color and texture via a green roof (*see Green Roofs, page 60*) or with free-standing containers.

The different roof surfacings have different textures, which dramatically affect the appearance of the roof. An aggregate surface roof can create a very uniform appearance. Conversely, every seam line and patched area will be apparent with some types of systems. The surfacing also affects the brightness, which can range from the low luster associated with common aggregate, to the glossy finishes on some metal panels.

Often the most visually distracting elements are rooftop mechanical units, boiler stacks, antennae and satellite dishes. These items should be placed to minimize their visual impact, or concealed by equipment screens when appearance is of concern. Mechanical equipment can be placed in penthouses, which not only offers visual relief, but more importantly enhances roof system service life by eliminating potential problems caused by penetrations and rooftop traffic to service the equipment. Also consider combining penetrations. For example, rather than designing four plumbing vents, combine the lines below the roof deck and penetrate the roof in only one area. With fewer penetrations, the appearance of the roof improves and the chance of leakage problems diminishes.

Paradoxically, some architects select a roof system for its visual attributes, while giving little consideration to the likelihood of the system actually being beautifully installed or remaining beautiful afterwards. For example, viewing a small sample of an attractive single-ply membrane does not reveal how objectionable seam lines or patches may be. Also, after months of exposure, the roof surface may look quite different than the office sample. This may be due to a weather-induced color change, or it may be due to accumulation of dirt in *birdbaths*, the small random depressions in the roof surface that allow collection of a small amount of water.

In misguided efforts to make the building look good, architects sometimes compromise functional aspects of the roof. A common example is failure to design copings for the tops of parapets.

Aesthetic considerations usually don't enter into the roof system selection process. But in those instances where aesthetic considerations are important, aesthetics should not take precedence over other design considerations discussed below or the other system selection criteria discussed in a subsequent chapter. Otherwise, the good-looking roof may be at risk of failing prematurely.

GEOGRAPHICAL INFLUENCES

The building's location is often a major design factor. Specific key issues include the weather during application, the weather during the roof's life and construction logistics. If the roof will likely be installed during cold weather, for example, a system conducive to cold-weather application should be specified or specifications should be written that provide special cold-weather application requirements *(figure 122 on page 112)*. If the building is located in an area that is likely to experience frequent rainfall or moderate wind speeds during roof application (such as Alaska's Aleutian Chain), the architect should specify a system that is able to accommodate these conditions (in these two cases, metal roofing would be worthy of consideration).

The weather conditions at the site during the roof's life can also significantly influence the roof design. For example, a roof at a site with very high temperatures and high solar radiation, high winds, or damaging hail needs to be designed in response to these severe conditions. These specific examples are discussed later in this chapter.

Although an infrequent factor, in some instances logistics is an important design consideration. A building in a remote location may economically limit the type of systems that can be used. In this case, systems that require little installation equipment are preferable (such as a torch- or cold-applied modified bitumen system). Logistics can also be a factor in urban areas. On a high-rise building, for example, a system composed of components that can easily be transported to the roof is desirable.

Some roof systems have characteristics that can cause adverse effects during application. Sprayed polyurethane foam may not be a good choice, for example, in a windy area adjacent to a building with windows or a parking area with cars because over-spray may be deposited on the glass. It may be possible to alleviate the problem by placing windscreens at the work area, by protecting the adjacent building's windows or by moving the cars. Or it may be best to consider an alternative system. Another undesirable characteristic is odor associated with BUR and with hot-applied modified bitumen systems. Some people find the odor to be objectionable up to a few hundred feet from the work site. This should be considered when designing a roof near buildings with sensitive occupancies such as hospitals, and when reroofing occupied buildings. To alleviate the asphalt odor problem, a torch- or cold-applied modified bitumen system or one of the other types of membranes discussed earlier in the chapter on "Membrane Materials" could be specified. Odor-control asphalt kettles are also available, which decrease the odors associated with BUR or hot-applied modified bitumen, but some odor is still released when the asphalt is applied onto the roof.

BUILDING CODES

The architect should determine if a building code has been adopted for the locale where the roof will be installed and, if so, what edition of the code is to be used. The last several editions of the three model building codes and the *International Building Code (IBC)* have many specific provisions related to the roof system, including reroofing projects. The code requirements can be quite different from those of the manufacturer or Factory Mutual (*see next section*). Since the building code has the force of law, it is incumbent upon the architect to ensure that his or her design complies with it. If there is a problem with the roof and the roof design is found to be in non-compliance with the code, the architect can face a serious financial problem.

If the building occurs in an area that has not adopted a building code, it is prudent for the architect to voluntarily comply with the roofing-related provisions of the current edition of the *IBC*. If there are subsequent roofing problems and the code was voluntarily complied with, the architect should be able to demonstrate that a prudent standard of care was exercised (at least with respect to those issues addressed in the code).

The building code may have specific requirements regarding documentation to be provided to the building department. For example, the *IBC* requires specific information related to wind loads (section 1603.1.4 of the 2000 edition). The *IBC* also has construction document requirements in section 106.1.1 of the 2000 edition.

Most building departments possess little expertise related to roof systems. Architects, therefore, should not rely upon the building department to discover non-code compliance during their plan review. Also, most building department inspectors do not inspect the roof system. Those departments that do inspect the roof likely possess insufficient knowledge of all of the low-slope systems that they could encounter.

FACTORY MUTUAL (FM)

FM Global (FMG) is the new name of the Factory Mutual Insurance Company and its affiliates. One of FMG's affiliates, Factory Mutual Research (FMR), provides testing services, produces documents that can be used by architects, and develops test standards for construction products and systems. FMR evaluates roofing materials and systems for resistance to fire, wind, hail, water, foot traffic and corrosion. Roof assemblies and components are evaluated to establish acceptable levels of performance. Some documents and activities are under the auspices of FMG and others are under FMR.

Roof products and assemblies that pass various FMR tests are "Approved." Approvals are listed in FMR's *Approval Guide*, which is published annually and should be in the architect's office. A CD version of the *Approval Guide* is issued three times a year. Systems that are not in the *Approval Guide* may be "Accepted." These systems, which include ballasted single-plies, require specific review by FMG. On many reroofing projects, "Acceptance" will need to be sought because the existing roof deck does not comply with current FMR criteria.

To obtain an "Approved" roof assembly, it is necessary that all components (for example, deck, fasteners and insulation) be "Approved." Substitution of a non-approved component during design or application will result in non-approval, which can significantly increase the cost of insurance or jeopardize availability of insurance from FMG.

As part of the approval agreement with manufacturers, FMR periodically makes unannounced audits of manufacturing facilities and products. Approved products are marked or labeled as such.

FMG produces several *Loss Prevention Data Sheets* related to roofs that are occasionally updated, but not on a prescribed cycle. The roofing-related data sheets are available in a three-ring binder titled *Loss Prevention Data for Roofing Contractors* (a current copy should be in the architect's office). The *Data Sheets* contain information that architects should consult when they want their designs to comply with FMG criteria.

Compliance with FMR criteria is not required unless mandated by the building owner's insurer or unless the building owner or material manufacturer voluntarily decides to require compliance. However, even in those cases where compliance with FMR is not required, it is prudent for the architect to consider the FMR criteria. Where compliance with FMR is required or desired, the architect should prepare the specifications and drawings so that they are in compliance with FMR requirements, rather than simply stating something like "comply with FM 1-49."

If the building is insured by FMG, the architect can seek FMG's assistance in answering questions related to interpretation of FMG or FMR publications, or seek "Acceptance" of materials or systems that are not "Approved."

FIRE PERFORMANCE

The roof assembly can be a major contributor to the spread of fire. It is therefore incumbent upon the architect to devote attention to this issue. Fire performance is addressed in the building code. Architects are cautioned that assemblies listed in the FMR and UL documents, discussed below, precisely describe the tested assemblies. Designs that deviate from the tested assembly (either with material substitutions or changes in thickness or arrangement) may adversely affect the fire performance of the assembly.

In establishing requirements for roof assemblies, three different conditions are considered: external fire resistance, internal fire spread and internal fire resistance.

External Fire Resistance

This condition considers the ability of the roof assembly to resist burning brands from adjacent buildings or forest fires. Roof assemblies are tested in accordance with ASTM E 108 (which is essentially the same as UL 790). Assemblies are rated as Class A, Class B or Class C. Class A has the greatest resistance; Class C offers very little resistance. The roof slope can affect the rating. For example, an assembly may have a Class A on a low slope, but only a Class C on a steep slope. Architects need to be aware of slope limitations related to the rating. External fire resistance is addressed in chapter 15 of the *IBC*.

Although Class A has the greatest resistance, some Class A assemblies are more resistant than others. Enhanced resistance may be desired in some instances, such as a building in an area prone to wildfires. This can be provided by a metal roofing system, as its roof surface is noncombustible. However, because heat can be radiated downward, it is important that combustible roof assembly components not occur below the panels, or if they do that a thermal barrier be provided between the combustibles and the metal roof components. Another option is to design a protected membrane roof with heavyweight concrete pavers. If the roof has parapets, enhanced parapet protection can be provided by installing mortar-faced extruded polystyrene boards over the parapet wall *(figure 30)*. However, the magnitude of the enhanced performance of this detail has not been quantified.

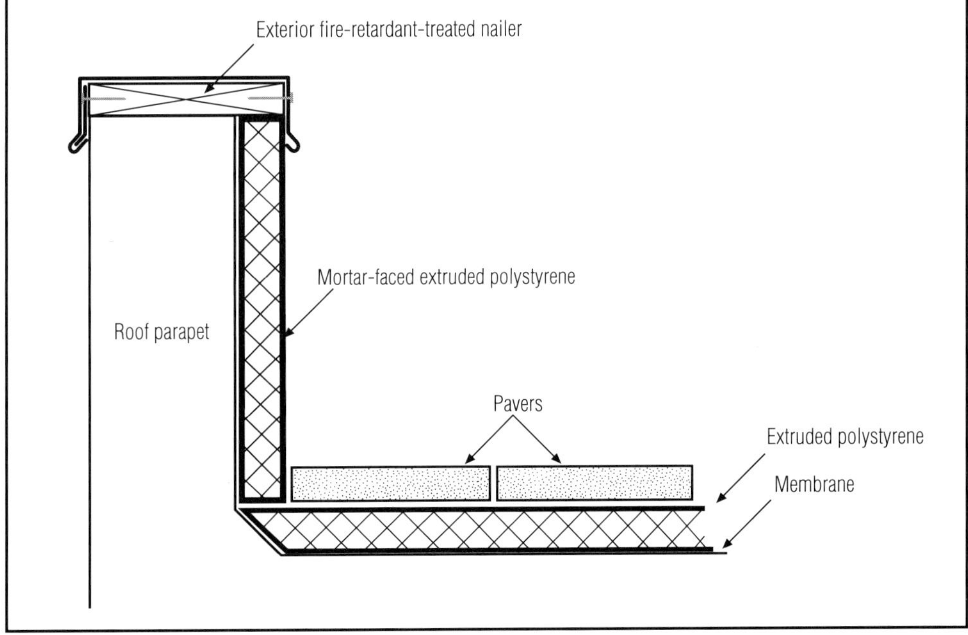

FIGURE 30
Enhanced Fire and Hail Protection
Extra protection is provided by installing mortar-faced extruded polystyrene over base flashing.

The UL *Roofing Materials and Systems Directory* and the FMR *Approval Guide* provide information on assemblies that were evaluated for external fire resistance.

Internal Fire Spread

This condition addresses spread of fire by fuel contribution from the roof system. In this scenario, a localized fire within a building liberates fuel from the roof system into the interior of the building where it propagates an under-deck fire spread. UL evaluates internal fire spread in accordance with test method UL 1256, and FMG uses test method FMR 4450. FMR uses three classifications—noncombustible, Class 1 and Class 2—to describe internal fire spread hazard. Noncombustible assemblies are the least hazardous, although most of the FMR approved assemblies are Class 1.

FMR recommends noncombustible or Class 1 assemblies. If an existing roof deck is Class 2, and if the building is FMG-insured or if the architect desires to comply with FMR recommendations, the architect should refer to FMG *Loss Prevention Data Sheet 1-29.*

Internal fire spread is addressed in chapter 26 (section 2603.4.1.5 of the 2000 edition) of the *IBC.*

The UL *Roof Deck Constructions* and the FMR *Approval Guide* provide information on assemblies that were evaluated for internal fire spread.

Internal Fire Resistance

This condition describes the resistance of the roof assembly and supporting beams and joists to upward passage of heat and flame, and resistance of the assembly and support structure to collapse because of elevated temperatures due to a severe internal fire. Roof-ceiling assemblies and support structure are tested in accordance with ASTM E 119 (which is essentially the same as UL 263). Roof-ceiling assemblies receive an hourly rating, such as 1½ hour or 1 hour, or they are non-rated. Resistance requirements are a function of types of construction as defined in chapter 6 of the *IBC.*

The type of materials used in the roof system as well as its thermal efficiency can affect the hourly rating. It is incumbent upon the architect, therefore, to specify no more insulation than that allowed by the tested assembly. This is particularly important in reroofing design, as the original assembly may have been based on tests that used significantly less insulation. If substantially more insulation is specified for the new roof, the hourly rating could be inadvertently downgraded.

The UL *Fire Resistance Directory Volume 1* and the FMR *Specifications Tested Products Guide* provide information on assemblies that were evaluated for internal fire resistance.

HAIL RESISTANCE

Heavy rain typically accompanies hailstorms. Hence, when hail damages a roof membrane, extensive interior water damage commonly occurs. The ensuing interior damage and business interruption can greatly exceed the cost of roof damage.

The *IBC* does not specifically address hail resistance of roof systems, but chapter 15 includes impact resistance requirements. FMG *Loss Prevention Data Sheets 1-29* and *1-34* do address hail. Both *Data Sheets* include a map of the United States, which shows the hailstorm hazard area as defined by FMG.

In the hazard area, roof systems meeting FMG's severe hail (SH) criteria should be used. Outside of the severe hazard area, FMG applies the moderate hail (MH) criteria. The hail resistance test method currently used by FMG has two limitations.

First, although the membrane is artificially aged prior to testing, the aging regime may not be sufficient for some types of membrane. Hence, a system may obtain an SH rating, but may become embrittled and very susceptible to hail damage after a few years of exposure. Second, the test is performed at room temperature. But during a hailstorm, the shattered hailstones cool the roof surface. Some membranes may resist hail damage while warm, but become susceptible to damage as their temperature decreases and they become brittle.

Another hail resistance test method is UL 2218. As with the FMG test, the UL test is performed at room temperature. However, the UL 2218 test does not artificially age the roof specimen before testing. Products passing the UL 2218 receive a rating of Class 1 to 4, with Class 4 providing the highest tested resistance to impact energy.

Architects are cautioned that designs that deviate from the FMG or UL tested assembly (either with material substitutions or change in thickness or arrangement) may adversely affect the hail performance of the assembly.

The hail delivered in some storms is quite large and very damaging *(see figure 118 on page 105)*. If hail resistance in excess of the SH or Class 4 rating is desired, a protected membrane roof with heavyweight concrete pavers is a good option. Heavyweight pavers are likely to be less prone to damage than lightweight pavers or mortar-faced extruded polystyrene boards. Alternatively, in lieu of insulation above the membrane, insulation could be placed under the membrane and a stone protection mat and pavers could be placed over the membrane. If the roof has parapets, parapet protection should be provided with, for example, mortar-faced extruded polystyrene boards *(see figure 30 on page 33)*.

Sprayed polyurethane foam roofs are also very effective in accommodating severe hail (Dupuis 1998). Although the coating and upper portion of the foam can be ruptured, the roof is not in immediate danger of leaking *(figure 31)*. A hail-damaged SPF roof can go unrepaired for many months without significant consequence. In most instances, repairs can be accomplished with the application of sealant at the impact areas.

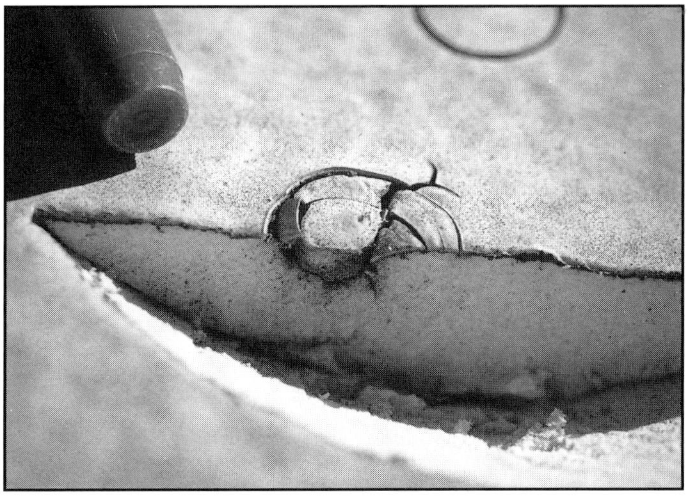

FIGURE 31

This is a view of a slit sample through an area of a SPF roof that was impacted by hail. The coating had ruptured. An electrical capacitance meter did not detect moisture in this area. The end of an ink pen shows the scale.

Another option to enhanced protection is to provide a secondary membrane, as discussed in Essential Facilities (*page 62*). With this approach, the primary membrane is sacrificed. However, the secondary membrane eliminates costly water damage on the interior.

WIND PERFORMANCE

When a roof blow-off occurs *(figure 32)*, the cost of interior damage and business interruption is typically far greater than the cost of damage to the roof itself . Also, the wind-blown roofing debris can cause damage to other buildings *(figure 33)* and personal injury or death. The *IBC* mandates that wind uplift loads be calculated for roof assemblies (see its chapters 15 and 16). The loads are to be determined in accordance with *ASCE 7*.

FIGURE 32

Most of the fully adhered single-ply membrane on the lower roof blew off, and much of the metal roof on the upper roof was also lost.

FIGURE 33

The metal roof-covering from another building penetrated the roof deck of this building.

Uplift loads can also be determined from FMG *Data Sheets*. In some instances, the loads derived from *ASCE 7* exceed those derived from FMG, but in other cases the FMG loads are higher. If the building is FMG insured, and the FMG-derived loads are higher than those derived from the building code, the FMG loads should govern. However, if the code-derived loads are higher than those from FMG, the code-derived loads should govern. The magnitude of difference between the code-derived loads and the FMG-derived loads can be extremely significant.

If a code other than the *IBC* is used, it may use a procedure other than *ASCE 7*. If another code is used, it is prudent for the architect to calculate the loads based on the code and the current edition of *ASCE 7* and use the loads from whichever procedure results in the highest loads. If the building is FMG insured, and the FMG loads are higher than loads derived from these other two procedures, the FMG loads should govern.

The analytical procedure used to calculate loads is based on rectangular shaped buildings. Buildings having unusual shapes, such as domes, require wind tunnel testing to determine appropriate pressure coefficients, or require the use of recognized literature documenting wind-load effects on a building similar to the one being designed.

The *IBC* requires load resistance of the roof assembly to be evaluated by one of the test methods listed in its chapter 15. For metal panel systems, the *IBC* requires test method UL 580 or ASTM E 1592. The American Iron and Steel Institute, however, has stipulated the use of E 1592. It is prudent for architects to specify use of E 1592 because it is more likely to give a better representation of the system's uplift performance capability. Architects are cautioned that designs that deviate from the tested assembly (either with material substitutions or change in thickness or arrangement) may adversely affect the wind performance of the assembly.

The *IBC* does not specify the magnitude of the safety factor that should be applied to the tested resistance. However, FMR uses a safety factor of two, which in most cases is a reasonable value.

Specifying load resistance is commonly done by specifying an FMR rating, such as FM 1-75. The first number ("1") indicates that the roof assembly passed the FMR tests for a Class 1 fire rating. The second number ("75") indicates the uplift resistance, in pounds per square foot, that the assembly achieved during testing. Applying a safety factor of two to this example, this assembly would be suitable where the design uplift load is 37.5 psf [1.8 kPa].

Because of building aerodynamics, or building-wind interaction, the highest uplift loads occur at the roof corners. The perimeter has a somewhat lower load; the field of the roof has the lowest load. FMG *Data Sheets* are formatted so that a roof assembly can be selected for the field of the roof. That assembly is then adjusted to meet the higher loads in the perimeters and corners by increasing the number of fasteners or decreasing the spacing of adhesive ribbons by a required amount. However, this assumes that the failure is the result of the pulling-out of the fastener from the deck, or that failure is in the vicinity of the fastener plate, which may not be the case. Also, the increased number of fasteners required by FMR may not be sufficient to comply with the perimeter and corner loads derived from the building code. Therefore, if FMR resistance data are specified, it is prudent for the architect to separately specify the resistance for the field of the roof (1-75 in the example above), the perimeter (1-130) and the corner (1-190).

For ballasted single-ply systems, the 2000 edition of *IBC* requires compliance with *Wind Design Standard for Ballasted Single-ply Roofing Systems,* also known as *ANSI/SPRI RP-4.* Unfortunately, it references the 1988 edition, rather than the 1997 edition. The 1997 edition of *RP-4* is more conservative and is therefore more appropriate to use. The 1997 edition of *RP-4,* however, still contains some non-conservative provisions. In high-wind areas, architects are encouraged to consider the recommendations derived from Hurricane Hugo research (Smith et al. 1992). For FMG-insured buildings, *Data Sheet 1-29* should be consulted.

Roof membrane blow-off is almost always a result of lifting and peeling of the metal edge flashing or coping, which serve to clamp down the membrane at the roof edge *(figure 34).* Therefore, it is important for the architect to carefully consider the design of metal edge flashings, copings and the nailers to which they are

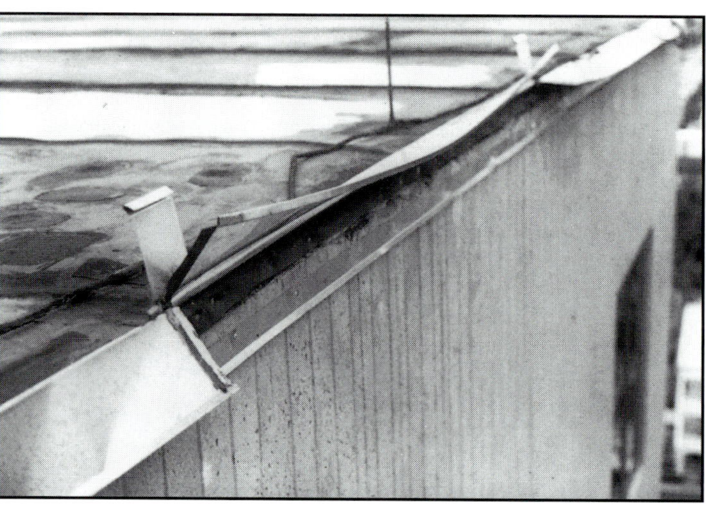

FIGURE 34

This metal edge flashing and cleat are 24 gauge [0.61 mm] stainless steel. Although this is a stiff flashing, it deflected and disengaged from the cleat. Fortunately, because this portion of the roof was shielded by a penthouse, the membrane did not lift and peel as is usually the case when the flashing disengages from the cleat.

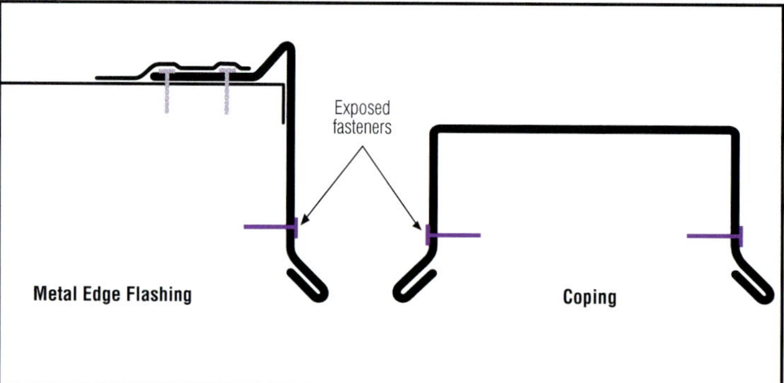

FIGURE 35

Face-fastening of Metal Edge Flashings and Copings

FIGURE 36

This single-ply membrane was punctured in several locations by missiles during a hurricane.

attached. *Wind Design Standard for Edge Systems Used with Low-Slope Roofing Systems,* also known as *ANSI/SPRI ES-1,* provides general design guidance, including a methodology for determining the outward-acting load on the vertical flange of the flashing/coping. (*ASCE 7* does not provide this guidance.) *ES-1* also includes test methods for assessing flashing/coping resistance. For FMG-insured buildings, FMR approved flashing should be used and *Data Sheet 1-49* should also be consulted. The traditional edge flashing/coping attachment method relies on concealed cleats that can deform under wind load and lead to disengagement of the flashing/coping. Storm-damage research has revealed that in lieu of cleat attachment, face-fastening of the vertical flange *(figure 35)* is a very effective and reliable attachment method (FEMA 1998).

Buildings located in hurricane-prone regions should receive special design attention because of the unique characteristics of this type of windstorm. In addition to being capable of delivering very high winds, hurricanes can deliver the strong winds for many hours, which can eventually lead to fatigue failure. The

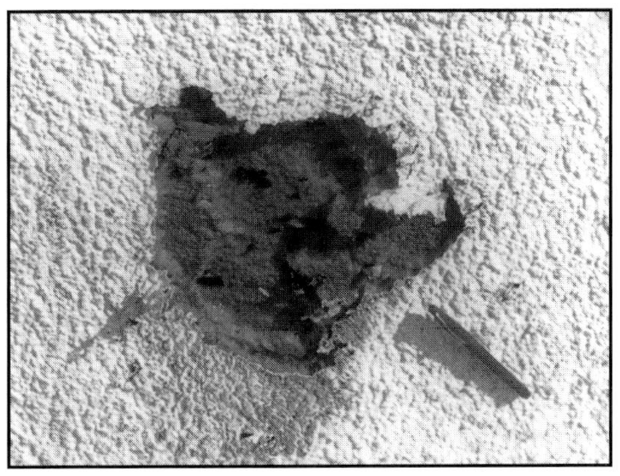

FIGURE 37
This SPF roof was gouged by a missile. However, because the missile did not completely penetrate the foam, the roof is still in a watertight condition and will remain so for a substantial period of time.

FIGURE 38
The flat portion of the roof is a fully adhered single-ply. The sloped portion is mechanically attached. A missile hit and tore the fully adhered area, and then tore into the mechanically attached membrane. The tear propagated in the reinforced mechanically attached membrane, resulting in a large failure.

FIGURE 39
An air retarder is being constructed by taping the joints of the plywood roof deck and sealing the deck/parapet interface.

direction of the wind can also change, thereby increasing the probability that the wind will approach the building at the most critical angle. Hurricanes also typically generate a large amount of missiles, or wind-borne debris, which can be very damaging to the roof *(figure 36)*.

Storm-damage research has shown that sprayed poly-urethane foam and liquid-applied roof systems are very reliable high-wind performers (Smith 1993, FEMA 1998). If the substrate to which the foam or liquid-applied membrane was applied does not lift, it is highly unlikely that the SPF or the liquid-applied membrane will blow-off. Both systems are also more tolerant of missiles than other systems *(figure 37)*. BUR and modified bitumen systems have also demonstrated good wind performance provided the edge flashing/coping does not fail (unfortunately, edge flashing/coping failure is common). Modified bitumen adhered to a concrete deck has demonstrated excellent resistance to progressive peeling after blow-off of the metal edge flashing. Metal panel performance is highly variable. Some systems are very wind resistant, while others are quite vulnerable. Of the single-ply attachment methods, the paver-ballasted and fully adhered methods are the least problematic. Single-ply membranes are very vulnerable to missiles, unless used in a PMR configuration or ballasted with pavers.

For recommendations related to mechanically attached *(figure 38)* and air-pressure equalized single-ply systems, see "Improving Wind Performance of Mechanically Attached Single-ply Membrane Roof Systems: Lessons from Hurricane Andrew" by Thomas Smith. This reference includes discussion of air retarders *(figure 39)*, which can be effective in reducing membrane flutter. Air retarders can also be beneficial in ballasted single-ply systems. This references also includes guidance on a detail that provides secondary protection against membrane lifting and peeling in the event that the edge flashing/coping fails: a bar can be placed over modified bitumen and single-ply membranes *(figure 40)*.

For recommendations related to metal panels, see "Insights on Metal Roof Performance in High-wind Regions" by Thomas Smith. For recommendations related to essential facilities, see the Essential Facilities section (*page 62*).

For a general introductory treatise on wind design, see the AIA's *Buildings at Risk: Wind Design Basics for Practicing Architects.*

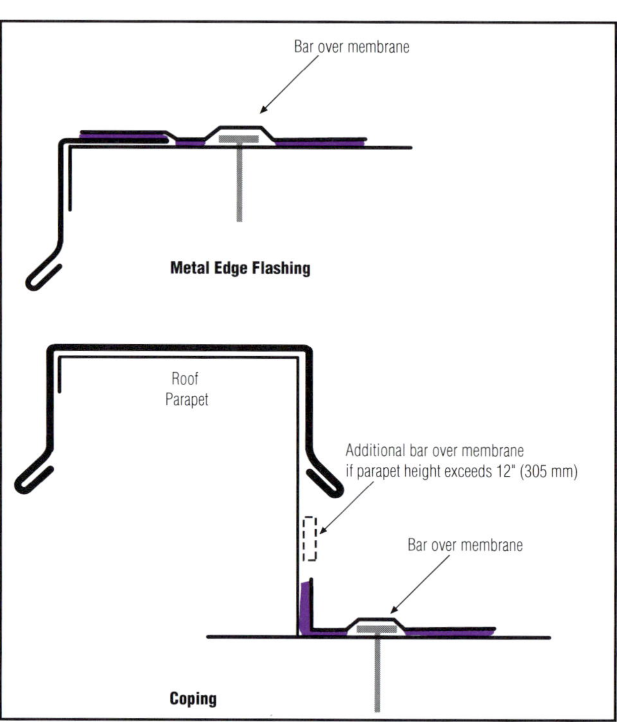

FIGURE 40
Bar over membrane near the edge of flashing/coping.

FIGURE 41
Almost the entire roof ponds water. The drain is located at a high point near the center of the photo. If a leak develops within the ponded area, a considerable amount of water can enter the building.

ROOF DRAINAGE

Slope is perhaps the greatest single design aspect that can significantly affect the performance of the roof system. Roofs that are dead level, or nearly so, present great opportunity for problems due to the large amount of water that can accumulate and stay on the roof for prolonged periods *(figure 41)*. In combination with solar radiation, ponded water can accelerate the degradation of field-applied coatings and some types of membranes. A minor membrane defect, such as a localized seam failure, may be of little consequence if there is adequate slope and the water simply flows downhill over the seam. But under an inch or two [25 to 50 mm] of ponded water, the same seam problem can allow a significant quantity of water to enter the roof.

For new construction, the IBC requires a minimum slope of ¼:12 (2 percent). For reroofing, a shallower slope is permitted as long as positive drainage is provided (the *IBC* definition of *positive drainage* is essentially the same as the NRCA definition, which is given in the "Glossary of Terms"). With the exception of metal roofs, providing slope greater than ¼:12 (2 percent) does not provide increased performance. While ½:12 (4 percent) minimum slope is recommended for metal roofs, increasing the slope of metal roofs beyond ½:12 (4 percent) is beneficial.

Slope can be developed by sloping the structure or by using tapered insulation. A sloped structure can be developed by sloping the deck support members, or in the case of a cast-in-place concrete or wet-fill deck, it can be accomplished by varying the deck thickness. Tapered rigid insulation is an expensive way to achieve slope. And each time the insulation is torn off during future reroofing, the expense of tapered insulation is incurred again.

When slope is achieved by sloping the deck support members, it is often difficult to provide adequate slope in all areas of the roof. Crickets or saddles made of tapered insulation may need to be specified in localized areas of the roof in order to achieve the desired drainage.

Whether achieving the slope with the structure or with tapered insulation, it is important to evaluate the roof with respect to deck deflection. *The NRCA Roofing and Waterproofing Manual* provides discussion on this issue.

The *IBC* references the *International Plumbing Code* for design of roof drainage systems. The water can be evacuated from the roof via internal roof drains, or by spilling water over the roof edge or through scuppers. For internally drained roofs, overflow drains or overflow scuppers are required. If overflow drains are used, they should be piped independently from the main drains. In establishing the placement of overflows, it is important to consider deflection characteristics of the roof structure. If deflection is not accounted for and a main drain is blocked, the ensuing pond may result in additional deck deflection and progressive pond growth that is not controlled by the overflows. This of course can lead to roof collapse. Drain lines should have an elbow or a bellows to permit roof deck deflection *(figure 42)*.

FIGURE 42
Roof Drain Line Routing

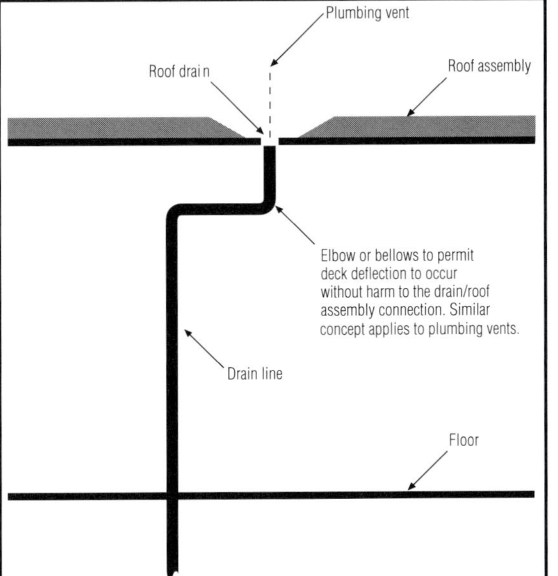

Spilling water over the edge or through scuppers also can present challenges. Scuppers should be large enough to avoid being clogged by debris or ice. The minimum size prescribed by the code to evacuate water may be susceptible to clogging.

In cold climates, spilling the water over the edge or through scuppers results in the accumulation of icicles. The falling ice may injure people or damage property *(figures 43 and 44)*. Even with a well-insulated and ventilated roof, icicle accumulation can usually occur. If snow is on the roof and it is a warm day, some snow will melt and refreeze to form icicles at the eave during the colder

ROOF DRAINAGE

FIGURE 43

This low-sloped metal roof drains over an eave, which is above entrances to stores. Ice, formed on the fascia, presented a falling hazard. The condition also created a slipping hazard because water dripping from the roof on warm days would freeze on the sidewalk at night.

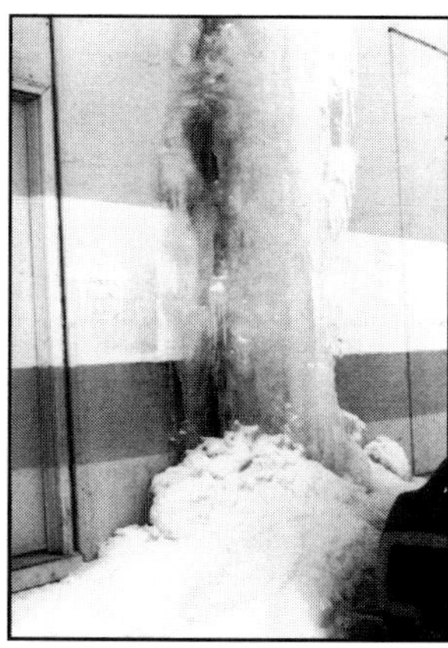

FIGURE 44

Ice accumulated on the wall of this two-story building between the scupper and the ground. On warm days, the ice fell into the parking area, creating a hazard. Although a sign on the wall says "Danger Falling Ice," a car is parked nearby.

FIGURE 45

Although this roof was well insulated (R-35) and it had a ventilation cavity between the insulation and metal panels, edge icing still occurred due to daytime snow melting and nighttime refreezing.

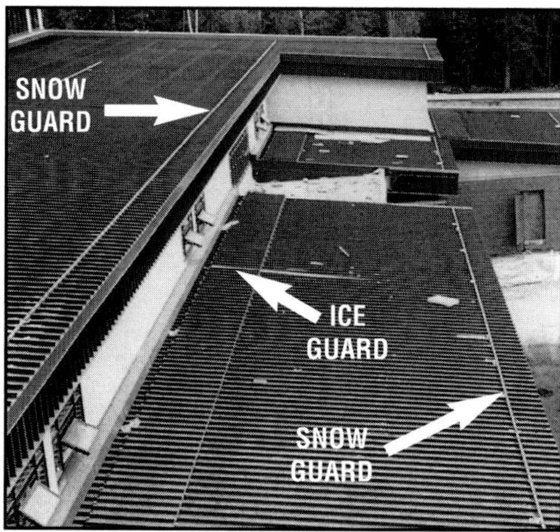

FIGURE 46

This school had a metal roof that was sloped at 1½:12 (13 percent). Ice falling from the upper roof could damage the lower roof. To prevent this problem, an ice guard detail was designed to protect the lower roof *(see figure 47)*. To prevent snow from sliding from the upper and lower roofs, snow guards were installed near the eaves *(see figure 49)*.

ROOF DRAINAGE

FIGURE 47

A 4-foot-wide [1219 mm] ice guard was placed under the drip line. The upper edge of the guard was 1 foot [305 mm] inboard of the drip line. The guard was made by placing preservative-treated plywood over neoprene sponge pads that were on top of the panel seams. Metal panels were placed over the plywood as a decorative surfacing. The guard was held in place with clips that allowed the guard to deflect when hit by falling ice. Without the guard, the metal roof would have been damaged.

FIGURE 48

A concrete paver-ballasted protected membrane roof was installed on this entry canopy. The pavers and XEPS insulation protected the membrane from falling ice. And the canopy prevented ice from falling into the entrance area.

FIGURE 49

A snow guard was placed near the edge of the roof to prevent snow slides. The guard should be well inboard of the eave so that expanding eave ice does not damage the guard. Debris needs to be periodically removed from the guard, particularly on school buildings.

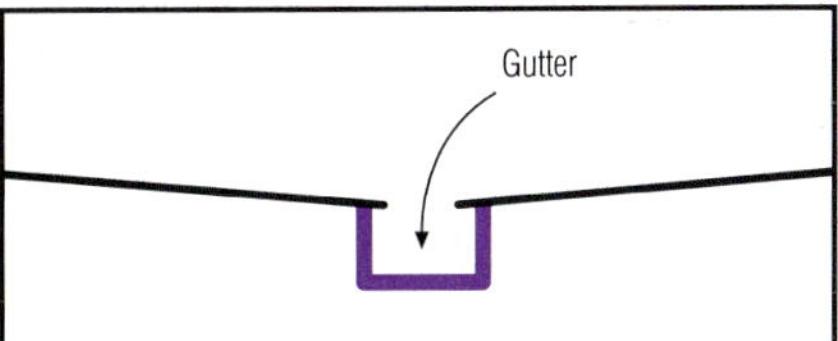

FIGURE 50

Avoid draining metal panel system into an interior gutter.

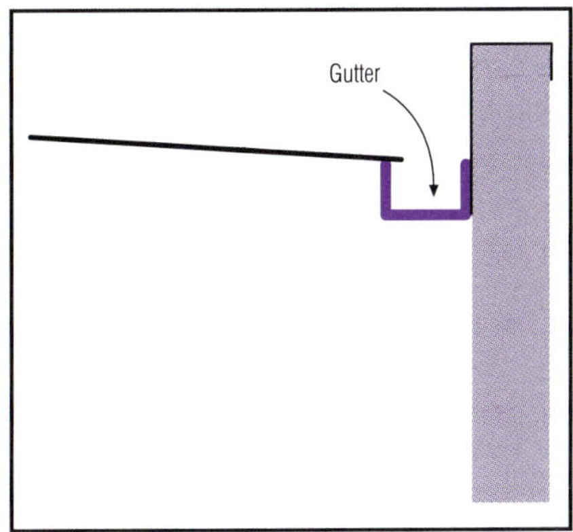

FIGURE 51

Avoid draining metal panel system into gutter at parapet.

evening *(figure 45)*. If the icicles can fall onto a lower roof, an ice guard can be designed to provide protection for the lower roof *(figures 46 and 47)*. If the icicles can form above a primary entry or exit from the building, a canopy *(figure 48)* can be design to protect people.

Sliding snow from slick-surface roofs such as metal panels and single-ply membranes can also injure people or block egress paths. Slides can be prevented on metal roofs with snow guards *(figures 46 and 49)*, or the snow may be permitted to slide if there is a safe zone for it to land. Snow slides can occur on roofs with very little slope. See Steep-Slope Roofs (*page 60*) for further discussion regarding snow slides associated with steep-slope roofs.

Metal roofs that drain into interior gutters (*figure 50)* or drain to gutters that are at the intersection of a parapet and sloped roof *(figure 51)* present challenging flashing problems. These conditions should be avoided, particularly in cold climates.

NCARB's *Low-Slope Roofing I* provides some further discussion on drainage issues.

DECK INFLUENCES

Of all the low-slope system options, structural metal roofing is the only one that does not require a roof deck. Structural metal panels have the capability of spanning between purlins. All of the other roof systems need a deck to support the insulation or the membrane itself. As the foundation for the roof system, a sufficient deck is critical to the success of the roof covering. NCARB's *Low-Slope Roofing I* discusses general deck design issues and the various types of decks that are available. *The NRCA Roofing and Water-proofing Manual* provides additional in-depth discussion.

If FMG compliance is required or desired, then the roof deck needs to comply with the FMR deck criteria.

General Design Considerations

Two general aspects of deck design are very important: construction-load carrying capacity and specifying decks that have high tolerance to moisture.

Construction-load carrying capacity

During construction, heavy equipment is frequently used, particularly on ballasted jobs and jobs where single-ply roll carriers are used *(figures 8 and 123 on pages 13 and 112)*. If the deck lacks sufficient strength to accommodate the construction loads (equipment or stored materials), deck fasteners (particularly welds) can be broken and the deck can be permanently deformed. A little extra money spent on strengthening the deck can be a very good investment. To provide more reliable steel deck attachment for both construction loads and wind uplift, screw-attachment rather than welding is recommended (Smith 1995c).

The architect should specify the maximum allowable construction equipment wheel loads and maximum allowable loads for materials stored on the roof.

Decks that tolerate high moisture

During the course of a building's life, it is likely that leakage will occur and the deck will be exposed to wet conditions. Decks that are capable of tolerating moisture for longer periods are desirable. Moisture-tolerant decks are less likely to require replacement, which can be quite costly and inconvenient to the building's occupants. And roof mechanics are less likely to fall through a deteriorated deck during reroofing *(figure 64 on page 52)*.

Cementitious wood-fiber decks are relatively susceptible to moisture degradation. The corrosion resistance of steel decks can be significantly increased by specifying a G-90 or aluminum-zinc alloy (Galvalume) coating in lieu of a prime coat of paint. Wood panel deck susceptibility to moisture can be decreased by specify preservative-treated panels.

System-specific Design Considerations

Design considerations for specific membrane systems are discussed below:

Mechanically attached single-plies

Because of the complex load distribution that occurs during wind loading of most mechanically attached single-plies, this type of system should only be considered when the deck is suitable. Twenty-two gauge [0.7 mm] minimum steel decks, wood plank, and 23/32 inch [18 mm] minimum thick plywood (not oriented strand board, or OSB) are typically good candidates (Smith 1995a). Low-density decks, such as lightweight insulating concrete and cementitious wood-fiber, lack resistance to the cyclical horizontal compression induced by the fasteners when they rock back and forth *(figures 129 and 130 on page 120)*. Cast-in-place concrete decks possess adequate compression, but drilling fasteners into this substrate is very expensive; another type of roof system, such as a ballasted single-ply or one of the several different types of fully adhered options, is generally more appropriate for this type of deck.

Fully adhered systems

Fully adhered modified bitumen and single-ply membranes are sometimes directly applied to the deck, but normally they are applied over rigid insulation boards. Depending upon the substrate, the boards are either mechanically attached or fully adhered. Fasteners securing insulation boards are not dynamically loaded, unlike those used for mechanically attached single-plies. Hence, mechanical fasteners can be used in low-density decks, provided a suitable type of fastener is specified.

During the last few decades of the twentieth century, asphalt was the predominate adhesive for adhesively attaching boards. However, use of other types of adhesives is re-emerging. One of the common alternatives is specially formulated polyurethane foam that is applied in ribbons *(figure 127 on page 118)*. Architects should be cautious of specifying adhesives other than asphalt, particularly in high-wind areas. After several years of experience, the industry may find that these alternative adhesives work well, or it may discover painfully that design, material or application improvements need to be made.

If the insulation boards are to be set in a continuous film of adhesive, there is less concern about their attachment than in those instances where the adhesive is discontinuous. Stresses are obviously higher in a discontinuous attachment. Some manufacturers promote adhesives for attachment of insulation boards to steel decks. Until there is ample long-term evidence that this is a viable approach, however, it is prudent for architects to specify that mechanical fasteners be used with this type of deck.

ENERGY EFFICIENCY

For buildings that have very large roof areas in comparison to their exterior wall areas, the thermal performance of the roof assembly is typically the dominant factor in the amount of energy consumed for heating and cooling the building. The thermal performance of the roof assembly is a function of the U-value of the assembly, the mass of the assembly and the reflectivity of the roof surface.

NCARB's *Low-Slope Roofing I* provides a general discussion on energy efficiency, the basics of heat transfer and heat flow calculations. NCARB's *Energy-Conscious Architecture, The NRCA Roofing and Waterproofing Manual* and *The NRCA Energy Manual,* which is included in *The NRCA Roofing and Waterproofing Manual,* provide additional in-depth discussion.

The architect should determine if the building code mandates a specified level of performance. For example, chapter 13 of the *IBC* requires compliance with the *International Energy Conservation Code.*

In addition the architect should be aware of the following:

RoofWise

A computer program, *RoofWise: The Digital Energy Workbench,* is available from NRCA. It can be used to determine thermal resistance requirements. It can also help determine the energy efficiency and approximate annual energy usage costs associated with various roof assembly designs.

As with most hand-calculation procedures, *RoofWise* does not account for the additional heat loss associated with insulation or membrane fasteners, nor does it account for the loss that occurs at insulation board joints. (Note: The smaller the board size, the greater the heat loss at joints. An advantage of a sprayed polyurethane foam system is that there are no board joints, nor are there mechanical fasteners. This is a highly energy-efficient system, particularly when surfaced with a reflective coating.)

If calculating these additional heat losses or gains is desired, see "A Heat Transfer Analysis of Metal Fasteners in Low-slope Roofs" by Douglas Burch for the fasteners and *A Methodology for Assessing the Thermal Performance of Low-Sloped Roofing Systems* by Walter Rossiter and R.G. Mathy for the board joints.

RoofWise also does not consider roof surface reflectivity.

Reflectivity

Reflectivity refers to the ability of the roof's surface to reflect solar energy. Reflectivity is primarily influenced by the color of the roof surface and, to a lesser degree, by the texture of the roof surface. The greater the reflectivity, the cooler the roof surface, which results in lower air-conditioning demand. Lower demand means reduced energy consumption and smaller, less expensive and more efficient HVAC equipment. Also, if the fresh-air intakes are located near the roof surface, a reflective roof offers another cost savings: The air stratum above a reflective roof will be somewhat cooler, hence the HVAC equipment has even less cooling to perform.

The Oak Ridge National Laboratory (ORNL) web site (www.ornl.gov/roofs+walls/facts/RadiationControl.htm) has a calculator for determining the economic benefit of reflective roof surfaces.

Highly reflective roofs also offer environmental benefits. In urban centers in warm climates, localized temperature excursions occur on sunny days because of heat-absorbing properties of the built environment. This phenomenon is referred to as *heat islands*. Dark-colored surfaces are currently a major contributor to the excursions. Not only do the temperature excursions place greater cooling demand on buildings, they also present environmental and health concerns. To meet the added cooling demand, more fossil fuels are consumed to produce electricity, which in turn increases power plant emissions that contribute to smog, acid rain and climate change. For further information about urban heat islands, visit the Lawrence Berkeley National Laboratory web site (http://eetd.lbl.gov/heatisland).

Reflectivity typically is somewhat reduced over time due to dirt accumulation on the roof surface. The roof surface can be periodically cleaned to increase its reflectivity as discussed below. The reflectivity of some roof surfaces also declines because of changes in physical properties of the roof membrane or coating. Most of the decline in reflectivity due to dirt accumulation or changes in physical properties occurs within the first three years after application. Further decline in the reflectivity after three years usually is insignificant.

To encourage the use of reflective products, the U.S. Environmental Protection Agency (EPA) developed the Energy Star® Roof Products Program. Products that meet EPA reflectivity criteria can carry an Energy Star® label. The criteria include meeting a specified minimum initial reflectance and meeting a specified minimum reflectance after three years of exposure. In reporting the three-year exposure value, manufacturers are permitted to clean the membrane prior to measuring the reflectance. Architects should therefore use discretion in using the three-year value in calculating energy savings. If the owner does not commit to periodic cleaning of the roof, the actual reflectivity value will likely be lower than that advertised by the manufacturer. Hence, the energy savings will be reduced. If this will be the case, the architect should attempt to obtain from the manufacturer the reflectivity value for the uncleaned membrane or coating after three years of exposure. The uncleaned value should be used to calculate heat gain and energy savings.

A highly reflective roof may also offer increased service life, because of the reduced membrane temperature. If the reflectivity is achieved by a field-applied coating, the coating may retard membrane aging.

Architects should not limit consideration of reflective roofs to warm climates. A roof in Fairbanks, Alaska, could be a good reflective roof candidate, where the summers are warm and sunny, and few buildings are air-conditioned. A reflective roof in that locale will help keep the interior temperature comfortable. Although heat-gain from dark-surface roofs is beneficial on cold sunny days, snow cover nullifies their advantage. For more information on this subject, see "Benefits of Reflective Roofs in Northern Areas of the U.S." by Thomas Smith.

For more information about the Energy Star® program, visit the EPA web site (www.epa.gov/appdstar/roofing/index.htm) and see "Energy Star® Label for Roofing Products, and a Contractor's View of the Energy Star® Label for Roofing Products" by Rachel S. Schmeltz and Timothy M. Davey.

INSULATION INFLUENCES

NCARB's *Low-Slope Roofing I* provides a general discussion on the types of insulation available. *The NRCA Roofing and Waterproofing Manual* provides additional in-depth discussion. The attachment of insulation boards to the deck was discussed in Deck Influences. When considering insulation choices during the design of a particular system, it is important for the architect to consider the following:

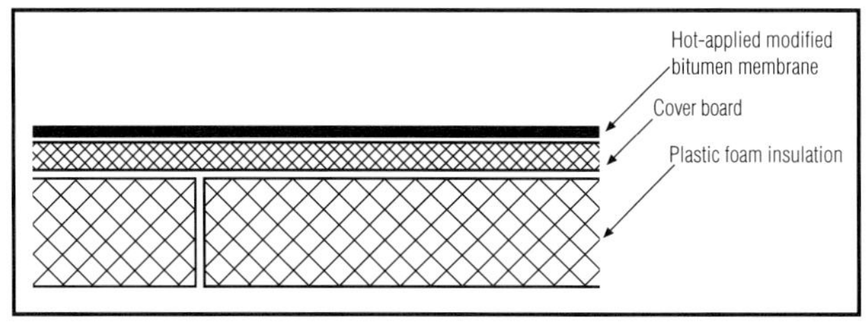

FIGURE 52
Hot-applied Modified Bitumen and Plastic Foam Insulation System

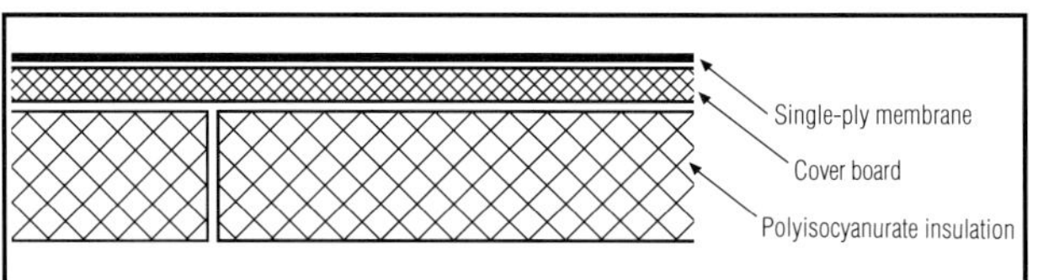

FIGURE 53
Fully Adhered Single-ply Over Polyisocyanurate

Fully Adhered Systems

When insulation boards are fully adhered, a maximum size of 4 feet x 4 feet [1219 mm x 1219 mm] is recommended. Also, the maximum thickness of each layer should be 2 inches [50 mm].

Verify that the adhesive between the insulation and the membrane is compatible with the insulation.

Modified bitumen membranes set in hot asphalt should not be placed directly over plastic foam insulation. Rather, NRCA recommends that a cover board such as perlite or wood fiberboard, or glass mat gypsum roof board (Dens-Deck) be adhered or mechanically attached to the plastic foam and the membrane mopped to the cover board *(figure 52).*

Although fully adhered single-ply membranes have commonly been placed directly over polyisocyanurate insulation, Bulletin 2000-3 from NRCA recommends placing a cover board over the polyisocyanurate (NRCA 2000). The cover board, such as perlite or wood fiberboard, or glass mat gypsum roof board, should be adhered or mechanically attached to the polyisocyanurate and the membrane adhered to the

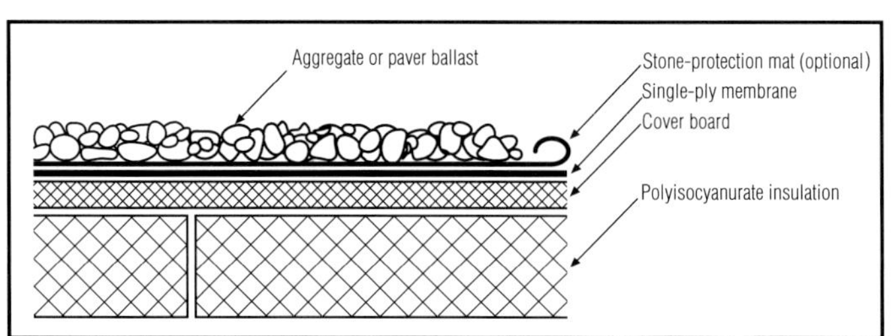

FIGURE 54
Ballasted Single-ply Over Polyisocyanurate

FIGURE 55
View of a polyisocyanurate insulation board after the aggregate-ballasted single-ply membrane was removed. The facer on the insulation is ruptured and the upper cells of the foam are crushed in many locations, which results in some R-value reduction. The small regularly spaced dots are factory perforations.

cover board *(figure 53)*. Without a cover board, rooftop traffic can cause the facer on the polyisocyanurate to delaminate. Once an area is delaminated, moderate winds can cause the delaminated area to propagate. The NRCA recommendation is prudent for many roofs. However, if a roof will be exposed to very limited traffic (such as a mechanical penthouse roof), there is little need for the cover board.

Ballasted Single-ply Membranes

The NRCA bulletin also recommends that a cover board be placed over polyisocyanurate insulation in ballasted systems *(figure 54)*. Ballasting work can cause localized crushing of upper portions of the insulation, resulting in some R-value reduction *(figure 55)*. The cover board protects the foam insulation during ballasting work. An alternative to installation of a cover board is to increase the thickness of the polyisocyanurate by a minimum of 0.1 inch [2.5 mm]. The additional R-value provided by the increased thickness should offset the loss due to crushing. This approach would be more economical than installing a cover board. For a discussion on rooftop traffic over ballasted systems, see Rooftop Traffic (*page 57*).

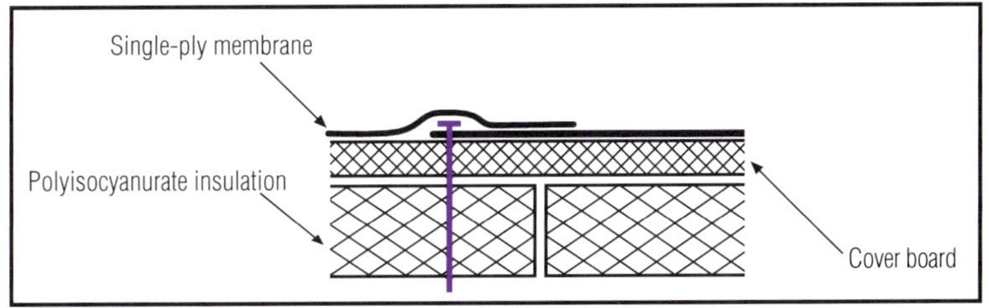

FIGURE 56
Mechanically Attached Single-ply Over Polyisocyanurate

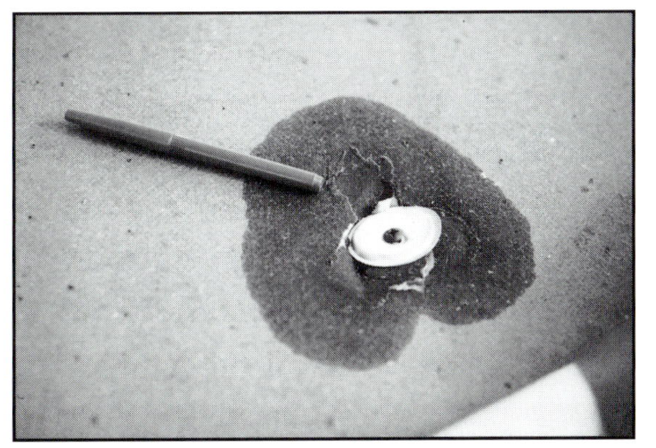

FIGURE 57
View of a screw and plate used to attach a mechanically attached single-ply membrane. The membrane blew off during a storm. Because this was a tab-attached system, the plate induced an eccentric load on the polyisocyanurate insulation. The compressive load induced by the plate exceeded the resistance of the foam; therefore, the foam was crushed. This resulted in loss of compression on the membrane between the plate and the insulation. With loss of compression, the uplift resistance was only provided by the tear resistance of the membrane in the vicinity of the screw. Because of high stress, the membrane tore around the screw shank and the membrane blew off.

Mechanically Attached Single-ply Membranes

NRCA Bulletin 2000-3 also recommends that a cover board be placed over polyisocyanurate insulation in mechanically attached systems *(figure 56)*. A cover board prevents the fastener plates from causing localized crushing of the polyisocyanurate during high winds *(figure 57)*.

For membranes installed over molded expanded polystyrene (MEPS) insulation, a minimum density of 1.35 pcf [22 kg/m^3] is recommended. This density is provided by boards complying with ASTM C 578, Type II or IX. For membranes installed over extruded expanded polystyrene (XEPS), a minimum density of 1.30

pcf [21 kg/m^3] is recommended. This density is provided by boards complying with ASTM C 578, Type X, IV, VI, VII or V.

Because of its limited resistance to compression, rigid fiberglass insulation should not be used in a mechanically attached single-ply system.

Polyisocyanurate Insulation

For polyisocyanurate insulation, a special report (NRCA 1999) recommended a minimum compressive strength of 23 psi [159 kPa]. This is substantially above the minimum compressive strength specified in the 2001 edition of ASTM C 1289. A Grade 3 product has a minimum compressive strength of 25 psi [172 kPa].

VAPOR RETARDERS

NCARB's *Low-Slope Roofing I* provides a general discussion on moisture flow, vapor retarders and calculation of the dew-point temperature. *The NRCA Roofing and Waterproofing Manual* provides additional in-depth discussion. With NRCA's *Roofwise* it is quick and easy to calculate the dew-point temperature and to confirm that the vapor retarder is properly located. (The retarder should be on the warm side of the dew point, and the retarder should be warmer than the dew-point temperature during winter design conditions.)

Vapor retarders are commonly associated with cold climates. A building with a very high interior humidity in a warm climate, however, may also need a vapor retarder. The 1996 edition of *The NRCA Roofing and Waterproofing Manual* was the first manual to provide a new methodology developed by the U.S. Army Cold Regions Research and Engineering Laboratory (CRREL) to assess the need for retarders in such conditions.

The most recent and sophisticated method for determining the need for a vapor retarder has been developed by ORNL. The ORNL procedure is based on computer simulation. It considers roof surface reflectance and moisture absorption properties of the roof insulation and deck, in addition to interior humidity and temperature and outdoor temperature. To use this web-based calculation procedure, go to the ORNL web site (www.ornl.gov/roofs+walls/) and click "interactive calculators" then click "moisture control in low-slope roofing." For further information on this

FIGURE 58
Do You Need a Vapor Retarder?

METHOD	CHICAGO LOCATION		KNOXVILLE LOCATION	
	Polyisocyanurate Insulation	Wood Fiberboard Insulation	Polyisocyanurate Insulation	Wood Fiberboard Insulation
NRCA	Yes	Yes	Yes	Yes
ASHRAE	Yes	Yes	Yes	Yes
CRREL	Yes	Yes	No	No
ORNL	Yes	No	Yes	No

The "NRCA" method refers to method that is based on interior humidity and exterior design temperature as defined in *The NRCA Roofing and Waterproofing Manual.* The "ASHRAE" and "CRREL" methods are discussed in *The NRCA Roofing and Waterproofing Manual. Table is courtesy of the Oak Ridge National Laboratory.*

FIGURE 59

Self-drying Roof Assembly
Most of the R-value is provided by the middle layer of insulation.

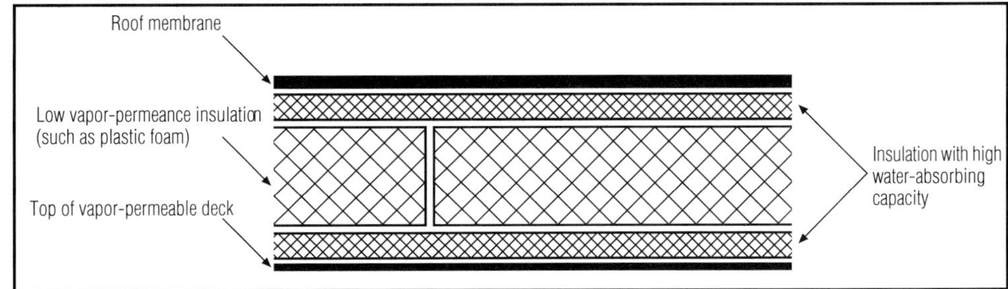

procedure, see "Review of Existing Criteria and Proposed Calculations for Determining the Need for Vapor Retarders" by A.O. Desjarlais and A.N. Karagiozis. The table in *figure 58* illustrates a comparison of the different methods for determining the need for a vapor retarder.

SELF-DRYING ROOFS

For buildings that don't need a vapor retarder, architects are encouraged to consider designing a self-drying roof *(figure 59).* A self-drying system allows for rapid downward dissipation of moisture (hence the roof deck needs to be somewhat permeable). It also needs to be composed of insulation materials that can accommodate some short-term accumulation of moisture without being damaged.

An advantage of a self-drying roof is that it can tolerate a moderate amount of roof leakage. After eliminating the leakage source (such as a puncture or small area of seam failure), the self-drying roof dries naturally and the system components retain their integrity. In contrast, wet roofing material would have to be removed if a similar leakage occurred on a roof system that is not moisture tolerant nor capable of self-drying.

The NRCA Roofing and Waterproofing Manual provides some discussion on self-drying roofs. For more in-depth discussion, see "Self-drying Roofs: What! No Dripping!" by A.O. Desjarlais.

FLASHINGS AND PENETRATIONS

NCARB's *Low-Slope Roofing I* provides extensive discussion on flashings and penetrations, including discussion on expansion and contraction considerations. Flashings and penetrations fall into three general categories:

- edge flashings, copings, base flashings and counter flashings
- flashings at plumbing, electrical and communications penetrations, and mechanical equipment
- flashings at expansion joints, seismic joints and area divider curbs.

It is incumbent upon the architect to design the flashing and penetration details and clearly illustrate and describe the requirements. Neglecting to give adequate direction on all flashing and penetration conditions on a roofing project is inviting failure of these important elements. *Figures 60 through 64* illustrate a few of the many types of flashing and penetration details that may be encountered.

It is important to extend base flashings a minimum of 8 inches [200 mm] above the membrane surface. By doing so, water inundation of the top of the base flashing is usually avoided. If the building is located in an area where slushy snow

FLASHING AND PENETRATION DETAILS

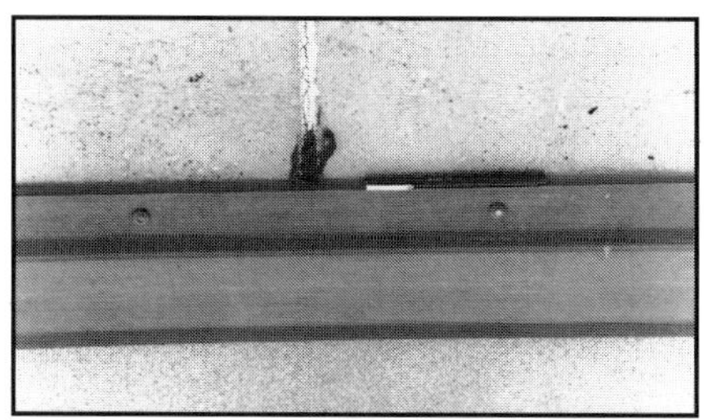

FIGURE 60

This surface-mounted counterflashing is on a precast panel. The counterflashing crosses over a vertical joint between panels. It is very difficult to achieve a long-term watertight connection at the juncture between the panel sealant joint and the sealant along the top of the counterflashing. This type of detail should be avoided. One option is to install furring strips over the precast panels and then install metal panels that overlap the counterflashing.

FIGURE 61

This metal roof is being installed in an area prone to daily rainfall. A special one-piece flashing assembly was factory-fabricated to accommodate four penetrations.

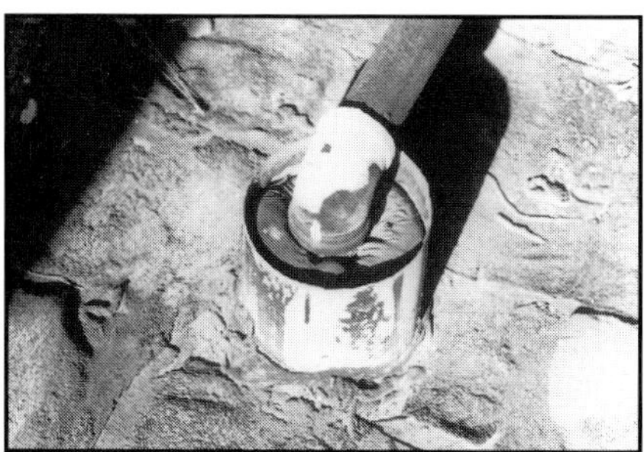

FIGURE 62

Pitch pockets, such as the one illustrated here, should be avoided whenever possible. The bituminous materials (traditionally used as a filler in pitch pockets) weather and become ineffective unless diligently maintained. An alternative to a pitch pocket is shown in *figure 63*.

FIGURE 63

This prefabricated curb cap can accommodate up to four penetrations. A flexible boot closes the opening between the curb cap and the pipe or conduit penetration.

FIGURE 64

This insulated BUR had been applied over a cementitious wood-fiber deck (which is susceptible to moisture problems) above a swimming pool in a cold climate. A curb and expansion joint had been installed above the deck. The cavity within the curb was not insulated nor was it protected with a vapor retarder. The deck was very deteriorated in the vicinity of the curb (it was possible to see the swimming pool below). Architects should pay attention to vapor control, and they should avoid specifying a deck that is relatively susceptible to moisture. Working on a deteriorated deck such as this one is potentially hazardous to roofing mechanics.

MECHANICAL AND ELECTRICAL EQUIPMENT DETAILS

FIGURE 65

This mechanical equipment is on an elevated stand that is sufficiently high to allow reroofing work to occur.

FIGURE 66

This mechanical equipment rests on a wood sleeper that is on a piece of cap sheet on the membrane. In high-wind areas, this type of detail is insufficient to prevent the equipment from being blown away. Also, when it is time to reroof, the equipment will need to be removed and reinstalled. Equipment should be on permanent curbs or stands.

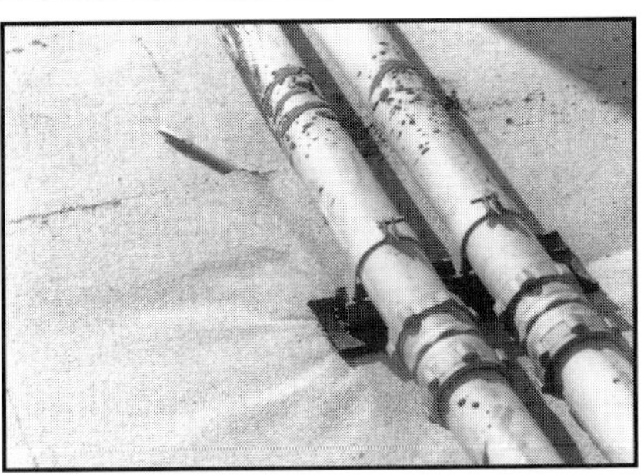

FIGURE 67

These electrical conduits were placed on a thin pad that rested on a modified bitumen membrane. The weight of the conduits crushed the insulation. The membrane will likely fail prematurely in the vicinity of the support. When the building is reroofed, significant extra expense will be incurred because the conduits will need to be lifted. If the conduits had to be on the roof, they should have been installed on elevated curbs so that reroofing could occur without having to move the conduits. Use of curbs would have also prevented the insulation damage.

FIGURE 68

This pipe is strapped to a block of wood that is on a piece of modified bitumen cap sheet. Thermal movements of the pipe caused the block to rotate and gouge the cap sheet. In addition, although the pipe is slightly elevated, it is not high enough to permit future reroofing. Pipes should be mounted on elevated curbs that permit expansion and contraction. Prefabricated units with rollers are available for this purpose.

EQUIPMENT BLOW-OFF CAUSED BY HIGH WINDS

FIGURE 69

This piece of equipment was on an adequately elevated stand. It blew off, however, because it was poorly attached.

FIGURE 70

The unit shown in *figure 69* was attached with four straps. The single screw between the strap and the unit had insufficient strength to resist the wind forces. A much stronger connection should have been designed for this location in southern Florida.

FIGURE 71

This mechanical equipment was blown off during a storm. Besides causing damage to the roof membrane, the displaced equipment left a big opening at the curb for rain to enter.

FIGURE 72

Two straps over this piece of equipment provided additional uplift resistance. However, the straps were thin and only attached with a single-screw. Straps on several other units were broken. This illustrates the importance of paying attention to the details when additional equipment securement is specified. Simply adding straps does not necessarily ensure performance.

FIGURE 73

This very large piece of equipment was blown several feet. It had clips to prevent horizontal movement, but there was no provision for uplift resistance other than its own weight. During very high winds, the weight of the equipment itself is insufficient to prevent blow-off.

STANDARD VERSUS ENHANCED DETAILS

FIGURE 74

This conduit penetration is too close to the wall. The conduit should be removed and placed at least 12 inches [305 mm] from the wall.

FIGURE 75

This rooftop is a maze of equipment. This presents a significant challenge to achieving a successful long-term roof.

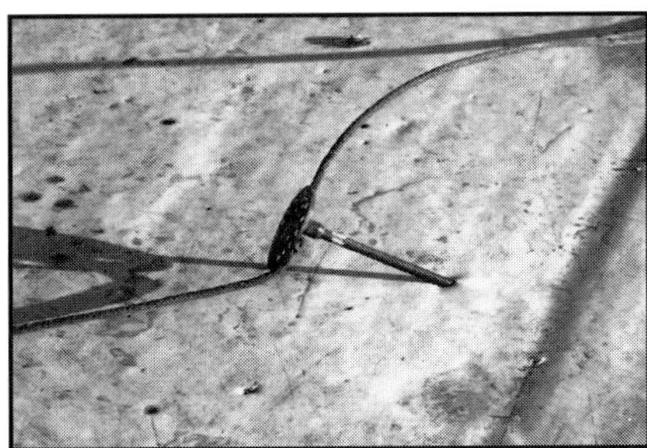

FIGURE 76

The attachment of air terminals, commonly referred to as *lightning rods*, and their cables typically receives very little attention. During a windstorm, this air terminal became dislodged and punctured the single-ply membrane in several locations. In high-wind areas, standard attachment details are insufficient to prevent the cable from dislodging and damaging the membrane.

FIGURE 77

Although it is common to place lightning rod cables directly on the membrane, it is prudent to place an extra strip of membrane underneath the cable. If water flows perpendicular to the cable, the strip should be periodically interrupted to permit water drainage. This photo shows a troublesome detail: the cable was not trimmed close to the splice clamp, so the cable pigtails are free to abrade the membrane.

FIGURE 78

In areas of high wind, fan cowlings or the entire fan unit frequently blow off. Cables were attached to this cowling and curb to avoid this problem. Keep in mind that the curb must be adequately attached or the curb and fan can be blown away.

FIGURE 79

View of an equipment curb before installation of the equipment. This is a PMR system. Flexible foam, as used for concrete expansion joint fillers, was placed between the curb base flashing and a sheet metal protective cover. Concrete pavers were placed over an "L" flange on the metal cover. This detail protects the base flashing from mechanical damage and direct weather exposure.

accumulates or exceptionally high winds accompany rainstorms, it is prudent to extend the height of the base flashings more than 8 inches [200 mm].

It is also important to adequately elevate stand-mounted mechanical equipment *(figure 65)*. Otherwise, future maintenance, repair and reroofing will be difficult or impossible unless the equipment is removed. *The NRCA Roofing and Waterproofing Manual* provides guidance on stand height, which is a function of equipment size. Mechanical and electrical equipment that is not stand-mounted should be placed on flashed curbs, rather than simply placed on sleepers that rest on the membrane *(figures 66 through 68)*. The attachment of the equipment to the stand or curb should be designed to resist the design wind loads *(figures 69 through 73)*. Unfortunately, the uplift attachment of mechanical and electrical equipment frequently is not addressed in the contract documents. This oversight often results in equipment blow-off during high winds.

Mechanical, plumbing, electrical and communications equipment should be located far enough away from nearby walls, parapets, edge flashings and other penetrations to allow the roof mechanic sufficient room to install the flashing *(figure 74)*. *The NRCA Roofing and Waterproofing Manual* provides spacing guidance. The amount of rooftop equipment should be minimized, and what equipment is on the roof should be laid out with thought so that the roof is not littered with equipment that presents an obstacle to successful roof system application *(figure 75)*.

Details of various conditions can be found in *The NRCA Roofing and Waterproofing Manual* (the details are also available on a CD-ROM for use with CADD), the *Architectural Sheet Metal Manual* by the Sheet Metal and Air Conditioning Contractors' National Association (SMACNA), and the *Metal Roofing Systems Design Manual* by MBMA. Standard details are also available from manufacturers. It is critical that standard details be evaluated for each project. In many instances, use of standard details will be appropriate. Standard details, however, are not always sufficient to provide reliable long-term service for certain climatic conditions or building characteristics *(figures 76 and 77)*. Enhanced details may be required in certain situations *(figures 78 and 79)*.

ROOFTOP TRAFFIC

Most roofs receive limited foot traffic. In these cases, special consideration to the roof membrane or surfacing is not necessary. However, on those roofs with frequent foot traffic (for example, to maintain rooftop equipment) or with window-washing equipment, the architect needs to consider the traffic.

One approach is to provide walkways. The other approach is to specify a roof system that has sufficient durability to withstand traffic throughout the roof. A problem with walkways is keeping workers confined to them. Instead of staying on the walkway, workers often walk on the field of the roof because the walkway does not lead directly to where they are going.

Walkways

A variety of walkway products are available, from metal grates and concrete pavers to various pads. Some are specifically designed for certain systems.

Aggregate ballasted systems are not comfortable to walk on. If traffic is frequent, installation of walkway pavers is recommended. If the membrane is over plastic foam insulation, the addition of a cover board over the foam in the vicinity of the paver can avoid damage to the foam by long-term foot traffic. If the cover board is just in the walkway vicinity, consideration needs to be given to drainage, so that the rise of the membrane at the cover board does not create a dam.

Metal roofs are also uncomfortable to walk on and foot traffic can abrade the panel finish. Metal grates are often used as walkways on metal roof systems. They are an effective solution in high-traffic areas.

FIGURE 80

A walkway pad was placed around this unit to provide protection during service. A wider pad should have been specified, however, to minimize the likelihood of items being placed beyond the pad.

On SPF systems, walkways can be made with additional layers of coating and granules.

In addition to specifying walkway protection, it is important to specify pads around mechanical equipment, since the roof is often damaged here by dropped tools or access panels *(figure 80)*. The base flashings should also be protected with pads.

More Durable Roof

Architects have several options if a tough roof surface is desired in lieu of walkways. The greatest resistance to rooftop traffic is provided by paver-ballasted systems, particularly in a protected membrane configuration, and mortar-faced extruded polystyrene boards in a PMR, both of which offer superb resistance. Next to this heavily armored approach is the modified bitumen system. Although not as tough as a PMR, it can accommodate traffic quite well. Surprisingly, sprayed polyurethane foam is also a good candidate. Although the coating and foam surface can be damaged by abusive traffic, the water-resistance of the foam prohibits the damage from causing serious problems. In high-traffic areas, re-foaming or re-coating damaged areas every few years can be an effective solution. A granule surfacing also provides greater resistance to traffic, and helps in providing greater traction when the surface is wet.

Exposed single-ply membranes are relatively susceptible to puncture and tearing. This is particularly true for non-reinforced membranes. Unless there is assurance that maintenance workers will not abuse the roof, it is prudent for architects to only specify reinforced membranes unless the membrane is in a PMR configuration.

Material Transport

Special handling procedures are required to avoid damaging the roof system if heavy equipment (such as HVAC units) needs to be transported across the roof after membrane installation. Placement by crane or helicopter should be specified for very heavy units. Units less than 1200 pounds [545 kg] can usually be transported by a cart with four large pneumatic tires. Rollers (pipes) should not be used to roll equipment across the roof. Equipment transport damage can be insidious. Damage may manifest itself soon after equipment is inappropriately moved across the roof *(figure 81)*. Or it may not be evident for several years: if the roof system damage is confined to a small localized area it may not sufficiently enlarge enough to be noticed until after years of exposure to thermal movements or wind-induced flutter.

FIGURE 81

This fully adhered single-ply membrane is no longer adhered. The damage was caused by transport of heavy equipment across the roof, which caused the insulation facer to separate from the cellular glass insulation. Moderate wind loads propagated the failure.

FIGURE 82

This roof in the light well presents special challenges. Because of the confined area, it is prudent to specify a system that emits little or no harmful or irritating vapors during application. Although workers can wear respirators, it is preferable to specify a system that can be applied without need to wear them. And in this case, with the large number of operable windows, there is also concern about vapors entering the building. A self-adhering modified bitumen membrane could be a good choice for this roof.

BUILDING SIZE AND SHAPE

In some instances, the building size or shape can be significant factors in system selection *(figure 82)*. For example, if a penthouse on a high-rise building is being reroofed, a system composed of materials and installation equipment that can be easily transported to the roof via the elevator and stairway is advantageous. Another example is a roof area that is only a few feet [about a meter or two] wide *(figure 83)*. This size of area does not

FIGURE 83
This small roof is a BUR. Many other systems would be much easier to apply on this small area.

readily lend itself to a BUR. In this instance, a modified bitumen (other than hot-applied) or a single-ply system may be the best choice.

CHEMICAL RESISTANCE AND COMPATIBILITY

Most roofs are not subjected to chemicals that are detrimental to them. But manufacturing facilities sometimes exhaust chemicals that may harm some types of membranes. Also, some membranes are harmed by grease exhausted from commercial kitchens. Fans can be specified that intercept the grease, but these are not always completely effective in prohibiting grease from reaching the membrane. Hence, in this example, specifying a membrane that is grease resistant would be prudent.

Architects also need to be aware of incompatibilities between system components. For example, PVC membranes are harmed when they are in contact with polystyrene insulation. The simple addition of a suitable separator sheet between the PVC and polystyrene insulation solves the problem.

INSECT ATTACK

In some locations (normally in the southern states), insects have attacked some single-ply membranes by boring holes through them. There is little published information about this issue. Architects designing single-ply roofs in locations where they have not practiced should consider contacting local roofing contractors to determine if insect attack is a problem.

CONSTRUCTION ABOVE OR ADJACENT TO A ROOF

Construction above a new or existing roof (for example, installation of cladding on a tower wall above a roofed base area), or construction of an addition adjacent to an existing roof, has the very great potential of causing harm to the new or existing roof. The other trades involved are unlikely to have respect for the roof; therefore, they can inadvertently cause damage.

Work Above a New Roof

If work will occur above a new roof, it is recommended that the architect specify one of the following options:

- Specify the installation of a temporary roof, which will be replaced with a permanent roof after the upper work is complete.
- Specify a paver-ballasted protected membrane roof.
- Specify a mortar-faced protected membrane system if only light abuse is anticipated. With this option, repair of the mortar facing will probably be required in a few areas.

Work Adjacent to Existing Roof

When constructing an addition adjacent to an existing roof, with limited service life remaining, it is prudent to specify that the existing roof be reroofed after completion of the addition. If the roof has ample life remaining, refer to the recommendations in the Reroofing Above or Adjacent to an Existing Roof section (*page 79*) in the following chapter. However, even when those recommendations are followed, saving an existing roof is challenging. In many cases it may be best to avoid the nondestructive evaluation (NDE) and protective measures and specify replacement after completion of the addition.

GREEN ROOFS

The term *green roof* was imported from Europe, where this roof system has been used for many decades. A green roof is one in which a layer of a special soil is placed over the roof membrane (or insulation over the membrane) and special types of plants are planted throughout the roof area. The system is popular in urban areas because of the visual attributes of the vegetation and because of the improved air quality provided by the plants.

There are two types of systems. The extensive system has a soil layer that is only a few inches deep [75 to 100 mm] and only supports small plants. The intensive system is about 3 feet [1 m] deep. It is used where trees and large shrubs are desired.

The design of a green roof requires special attention, as finding a leak and repairing it is expensive. Among several important factors, a membrane must be selected that is resistant to root penetration or a separate root-barrier material needs to be placed above the membrane.

In urban areas that have problems with rapid runoff after rainfall, green roofs are advantageous as a portion of the rainfall is retained within the soil. For further information on this aspect of green roofs, see "La retenue temporaire des eaux pluviales: une fonction economique et ecologique des toitures a faible pente" by A. Chaize.

A research project in Canada evaluated the effect of soil depth on winter damage to plants. It also evaluated a large number of plant species applicable to the North American climate. For more information on this, see "Extensive Green Roofs" by Marie-Anne Boivin.

STEEP-SLOPE ROOFS

All of the membranes discussed in the earlier chapter on "Membrane Materials" may be used on steep slopes. However, special considerations need to be made, including:

- Ballasted and protected membrane systems are limited to a maximum slope of 2:12 (17 percent), hence they are not applicable to steep slopes, which by definition are greater than 3:12 (25 percent).
- Where snow can accumulate, consideration needs to be given to sliding snow. Snow slides are typically not problematic with granule-surfaced membranes. However, damaging slides can occur on metal and single-

FIGURE 84

This barrel vault originally had a BUR with a mineral-surface cap sheet. It was reroofed with a fully adhered EPDM. Because the surface of the EPDM was much slicker than the cap sheet, an avalanche occurred. At the base of the barrel, there was a large gutter. Although the gutter was framed with wood and quite strong, the avalanche severely damaged it. Adjacent buildings that were identical in design to this one, but had retained the original BUR cap sheets or had new granule-surfaced modified bitumen roofs, did not experience snow slides.

ply roofs because of their slick surface (*figure 84*). Snow guards can effectively address the sliding problem on metal roofs (*figure 49 on page 43*) if the slope is not too great, but the problem is difficult to solve with single-plies.

- For modified bitumen systems, working with hot asphalt on steep slopes is potentially dangerous, and use of cold adhesive may lead to membrane slippage problems before the adhesive cures. Therefore, torching is usually the preferred option for steep slopes.

MAINTENANCE

Maintenance and repair demands will be directly influenced by decisions made by the architect. The frequency and costs of required maintenance and repair will be increased if less durable materials are specified, if a system is selected that does not have inherent long-term durability or if marginal details are designed. The building owner's willingness to commit to a suitable maintenance program should influence the architect's decision regarding system selection, as is discussed in the chapter on "System Selection Criteria."

If a coating is specified, the architect should advise the building owner that periodic re-coating will be required. Re-coating is particularly important if the coating is necessary for the roof assembly to provide the required fire performance. Also, if a reflective coating is specified to reduce energy consumption, re-coating will be periodically required to avoid a dramatic change in roof reflectivity as the coating weathers away. Re-coating will probably be required about every five to 15 years, depending upon climate, quality and type of coating, applied thickness and quality of application.

FIGURE 85

View of neoprene flashing at a plumbing vent on an older EPDM membrane. The neoprene flashings (which are no longer used in the industry) did not offer long life. If the roof is periodically inspected, this type of problem can be found and corrected before significant damage occurs.

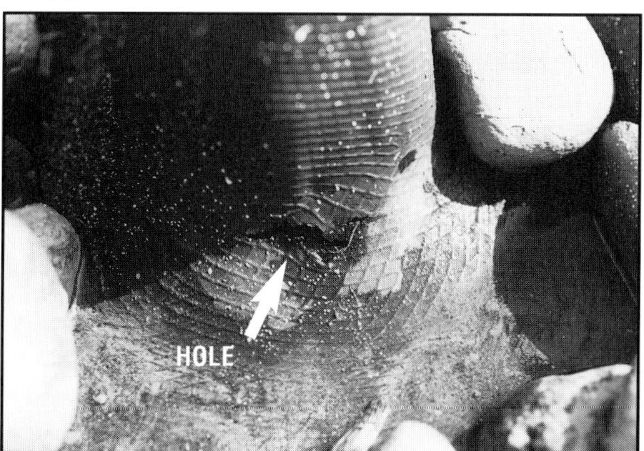

See Project Close-out (*page 121*) in the "Construction Contract Administration" chapter regarding owner maintenance manuals and semi-annual roof inspections. A building owner who commits to having the roof periodically inspected and maintained is more likely to become aware of minor problems that can be easily repaired (*figure 85*). However, minor problems that are not detected and corrected early often result in a premature end to the roof's life.

ESSENTIAL FACILITIES

FIGURE 86

This hospital lost its roof membrane during hurricanes in 1989 and 1995. Both times, the facility had to be evacuated and medical treatment had to be provided in tents. Because this was a tropical climate, the lack of air conditioning in the temporary facilities was greatly missed. Hospital roofs in hurricane-prone regions should have fail-safe roof systems.

Essential facilities are defined in *ASCE 7*. They include hospitals, police and fire stations, and evacuation shelters. Because of the importance of these buildings, essential facilities are designed for higher wind, snow and seismic loads. These buildings should also be designed with more robust, fail-safe roof systems *(figure 86)*, although presently this is not prescribed in the *IBC*.

Consider the following recommendations when designing roofs on essential facilities in hurricane-prone areas and in areas that frequently experience tornadoes (Smith and McDonald 1993). It is also prudent for architects to follow these recommendations when designing roofs for buildings that contain expensive equipment or critical operations (such as computer centers and research labs) in hurricane- and tornado-prone areas, even though they are not designated as essential facilities.

Deck

A deck of cast-in-place concrete is preferable to other types of decks because it typically provides very reliable resistance to wind uplift. Also, it is unlikely that missiles (wind-borne debris) will penetrate this type of decks, thus providing greater protection to occupants.

Membrane

FIGURE 87

The single-ply membrane on the hospital shown in *figure 86* was littered with roofing missiles, including an exhaust fan, continuous edge flashing cleats and a coping. The roof surface should be able to resist these types of missiles, or a secondary membrane should be specified to provide watertight protection.

Because of the great potential for missile damage to the membrane *(figure 87)*, specify one of the following options:

- Liquid-applied system over cast-in-place concrete
- Modified bitumen membrane torched directly to cast-in-place concrete
- Secondary membrane:
  Install a secondary membrane over the deck (depending upon the deck type, a cover board may be needed over the deck). Seal the secondary membrane at perimeters and penetrations. Specify a minimum of 2-inch [50-mm] thick rigid insulation over the secondary membrane to absorb missile energy. A cover board may be needed over the insulation,

FIGURE 88

This roof has been designed with a secondary membrane consisting of a torch-applied modified bitumen base sheet. Above this base sheet is a layer of polyisocyanurate insulation, then a layer of glass mat gypsum roof board and finally a modified bitumen membrane (which has not yet been installed).

depending upon the type of insulation and type of primary membrane. If the primary membrane is punctured during a storm, the secondary membrane should provide watertight protection unless the roof is hit with missiles of very high energy (*figure 88*). (Because of the difficulty in assuring watertight integrity at the secondary membrane, mechanically attached single-ply systems are not recommended for the primary membrane.)

- Sprayed polyurethane foam roof system: It is recommended that the foam be a minimum of 2-inches [50-mm] thick to avoid missiles penetrating through the entire layer of foam.
- Protected membrane roof with heavy concrete pavers: It is recommend that pavers weighing a minimum of 22 psf [107.4 kg/m^2] be specified. In addition, the base flashings should be protected as illustrated in *figure 30 on page 33,* except that a sheet metal cover should be placed over the mortar-facing. Parapets are recommended at roof edges. The parapet should be at least 3 feet [1 m] high; higher if so indicated by ANSI/SPRI RP-4. In extreme wind environments (for example, a very high basic wind speed coupled with a tall or unusually shaped building), a parapet higher than either 3 feet [1 m] or the height derived from RP-4 is recommended.

Metal Edge Flashings, Copings and Nailers

A safety factor of three is recommended for the design of metal edge flashings, copings and nailers. Compliance with *ANSI/SPRI ES-1* is also recommended. Use of exposed face-fasteners should be considered (*see Wind Performance section, page 36*).

Rooftop Equipment

A safety factor of three is recommended for the attachment design of rooftop equipment. Special consideration should be given to the attachment of fan cowlings *(figure 78 on page 55)* and equipment access doors, as these components are frequently blown off.

Aggregate and Pavers

Aggregate, lightweight pavers and standard weight pavers are not recommended because of the potential for blow-off *(figure 89).*

FIGURE 89
Aggregate ballast was blown off of this hospital during a hurricane. The aggregate broke a large number of patient room windows. And ambulances arriving at the hospital during the storm were pelted with aggregate.

Reroofing

For reroofing projects, it is recommended that the existing roof system be torn off rather than re-covered so that the deck integrity and attachment can be verified and strengthened if necessary (*see next chapter, page 66*).

DURABILITY

Durability refers to the capability of the roofing system to remain serviceable, or functional, over a minimum specified time. Durability is a function of the appropriateness and quality of the materials, the quality of the system design and detailing, the quality of application, and attention to repair and maintenance. Some systems are inherently more durable than others. But the durability of any of the low-slope systems can be increased by design enhancements. For example, specifying a stone-protection mat between the membrane and aggregate ballast greatly minimizes the likelihood of membrane puncture (as discussed in the "Membrane Materials" chapter), specifying a field-applied coating will increase the life of most APP modified bitumen membranes, and specifying a reinforced EPDM instead of a non-reinforced EPDM results in a system that is less likely to experience shrinkage-induced problems.

Selecting and designing a more robust system is prudent. Such a system is much more capable of accommodating severe weather, lack of maintenance, minor abuse, and less than perfect design, materials and workmanship. It is particularly important to select and design a durable and reliable roof assembly when the building is an essential facility or a building with very valuable contents or operations.

RELIABILITY

Reliability is sometimes thought of as being the same as durability, but it is not. Reliability refers to the performance that can be expected of a large number of similar roofs exposed to similar conditions. Within a large population of roofs, there will be those that are very poorly designed, installed or maintained, or were constructed with poor materials. There will also be those roofs that were designed, installed and maintained exceptionally well, and constructed with top-quality products. These two extreme groups of roofs will perform, respectively, much worse and much better than the majority of the roofs, which received average attention and were constructed of good quality materials. It is this average group of roofs that reflects the inherent reliability of the roof system.

Limited data are available to help discern the relative reliability of systems. However, in evaluating the likely demise of the system, it is often possible to identify reliable systems based on limited research, anecdotal information and rational evaluation. For example, in an area with a significant hail threat, it would make sense that a paver-ballasted protected membrane should be very reliable in comparison with an exposed membrane. And in areas with a very high wind threat, a liquid-applied system over cast-in-place concrete has been shown by research to be very reliable (FEMA 1998). It is prudent for architects with limited roofing expertise to obtain reliability input from a qualified roofing contractor or roof consultant.

ROOF SYSTEM WEIGHT

The table in *figure 90* provides an approximation of the weight of various types of roof coverings.

FIGURE 90
Roof Covering Weights (Approximate)

SYSTEM TYPE	WEIGHT[1]
BUR, smooth surface	1.5 [7.3]
BUR, aggregate surface[2]	8 [39]
Modified bitumen[3]	1 to 1.5 [4.9 to 7.3]
Single-ply, fully adhered or mechanically attached	0.5 [2.4]
Single-ply, aggregate-ballasted[4]	13.5 to 17.5 [65.9 to 85.4]
SPF, coated surface[5]	0.5 [2.4]
SPF, coated surface and granules[5]	0.7 [3.4]
Liquid-applied[6]	0.3 to 1.5 [1.5 to 7.3]
Metal panels[7]	1.5 to 2.5 [7.3 to 12.2]

NOTES:

1. The approximate weights, in psf [kg/m^2], are only for the membrane and membrane surfacing. The weight of the insulation is not included, except for the SPF systems.
2. BUR aggregate surface: A surfacing of 4 psf [20 kg/m^2] is normally specified for aggregate (slightly less for slag). However, the actual installed weight can exceed this value. Therefore, a value of 6 psf [29 kg/m^2] was used for the aggregate for design load purposes. The weight of a four-ply membrane was added to the 6 psf [29 kg/m^2] load to obtain the value in the table.
3. The modified bitumen value is for a two-ply cold-applied system with a mineral surface cap sheet.
4. Aggregate-ballasted single-ply: A surfacing of 10 psf [49 kg/m^2] is normally specified for #4 aggregate, and 13 psf [64 kg/m^2] for #2 aggregate (which is used in high-wind areas). However, the actual installed weight can exceed this value. Therefore, a value of 13 psf [64 kg/m^2] was used for the #4 aggregate and 17 psf [83 kg/m^2] for the #2 aggregate for design load purposes. The weight of the membrane was added to aggregate design load values to obtain the values in the table.
5. The SPF values include 1 inch [25 mm] thickness of sprayed polyurethane foam. A value of 0.25 psf [1.2 kg/m^2] was used for the foam itself.
6. The liquid-applied values are based on application directly to the roof deck.
7. The metal panel value is based on 22 gauge [0.7 mm] aluminum-zinc alloy steel panels. Weight depends on rib height and spacing.

REROOFING CONSIDERATIONS

Although architects are frequently involved in reroofing projects, many architects do not perform reroofing design work unless it is in conjunction with the design of an addition to an existing building. Reroofing design is substantially more complex than designing roofs for new buildings because of several reasons, including:

- because the building is finished and occupied, there is risk of water damage to the interior and in some cases there is risk of injury to occupants;
- existing conditions often present obstacles to good roofing practices (*figures 91 and 92*);

- unknown conditions, such as deteriorated deck or poorly attached edge flashing nailers (*figure 93*);

- the rare presence of friable-asbestos-containing fireproofing below the deck, on the underside of the membrane, or at flues.

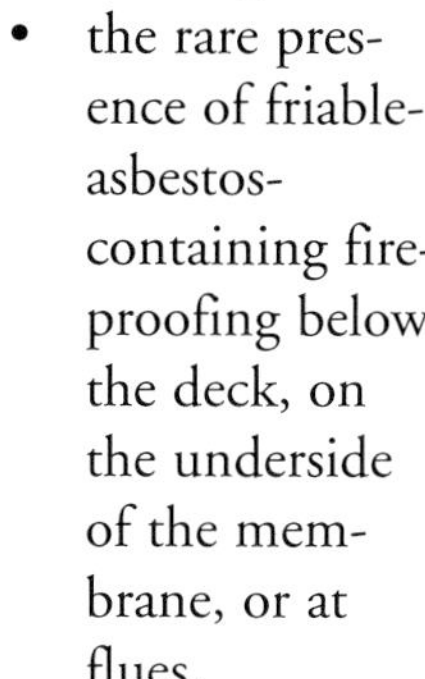

FIGURE 91

Casually accepting awkward conditions is not prudent. In this project, for example, a short wall divided an upper and lower roof. During reroofing, the modified bitumen membrane was butted to the metal wall panels. The panels should have been removed or skinned over with plywood. In addition, a proper cant, base flashing and counterflashing should have been installed. Although it would have cost more initially, this alternate approach likely would provide a much longer service life and a good return on the investment.

FIGURE 92

Although costly, low windowsills should typically be raised. This existing windowsill, for example, was very close to the roof surface. It was not raised prior to the application of the SPF system. This system does a good job in accommodating a low sill condition, provided there is sufficient room to install the foam without blocking weepage paths.

It is therefore imperative that architects involved in reroofing design be diligent and thorough in their evaluation of the existing roof system, and in the selection, specification and detailing of the new system. If the architect does not have considerable reroofing design expertise and experience, it is prudent to consult with a professional roofing contractor or roof consultant.

FIGURE 93

When this mechanically attached single-ply was installed during a reroofing project, the existing wood nailers for the metal edge flashing were reused, as is common. The nailer lifted up, however, because it was inadequately attached in this area. After the nailer lifted, the membrane was very vulnerable to blow-off.

Reroofing is defined as the process of re-covering or tearing off and replacing an existing roof system. *Re-covering* is defined as the addition of a new roof system over an existing roof system. *Tear-off* is defined as the removal of the existing roof system down to the roof deck and the installation of a new roof system over the existing deck. A *partial tear-off* is defined as tear-off of the existing roof system in selected areas of the roof.

There are two variations of a partial tear-off. In one case, the existing roof system is removed down to the deck in selected areas of the roof (typically those portions that have wet insulation). Infill materials are installed at the tear-off areas and a new roof system is added over the infill areas and over the area where the membrane was not removed. In the second case, the membrane is torn off throughout the roof area, but the insulation remains (unless there are wet areas, in which case, the wet insulation is also removed). A new roof system is added over the existing insulation.

Substantial costs can be saved by reroofing before the existing roof insulation gets wet (because of the opportunity to retain the insulation), or before interior finishes, furnishings or equipment are damaged or destroyed. Unfortunately most roofs are not periodically observed, hence most problems go unnoticed until they become quite severe and costly to correct. See the "Construction Contract Administration" chapter for a discussion about semi-annual observations.

Reroofing presents an opportunity for implementing improvements, such as enhanced thermal performance to reduce cooling or heating costs, improved drainage, and increased resistance to fire, snow or wind loads.

Reroofing design involves several other parameters in addition to the design considerations discussed in the previous chapter. The following topics are discussed below:

- evaluation of the existing roof assembly
- repair or reroof
- tear-off versus re-cover
- thermal upgrade
- temporary roof
- special reroofing design considerations
- reroofing above or adjacent to an existing roof
- special aspects of specifications and drawings
- special aspects of construction contract administration.

At the end of this chapter, a few publications are listed that provide further information on reroofing.

EVALUATION OF THE EXISTING ROOF ASSEMBLY

The first and most crucial phase of the reroofing design process is the evaluation of the existing roof assembly. The evaluation needs to be conducted by an investigator who is experienced in the type of roof system that is currently on the building. The purpose of the investigation is to determine the composition of the assembly, including the thickness of each type of insulation board within the assembly, the condition of the existing assembly and causes of failure. An aging failure is the desired failure mode. In this type of failure, the roof simply wears out after many years of exposure. However, if the roof is failing prematurely, it is important to determine the causes of the premature failure so that the new roof can be designed to avoid replicating the problem. During the investigation, potential problems unrelated to the roof need to be investigated, such as leakage from walls that extend above a roof or leakage from rooftop HVAC units. If non-roof related deficiencies are causing leakage but not addressed, roof replacement will not solve the problem.

The extent of the problems identified during the field investigation needs to be determined. For example, does wet insulation occur only in isolated areas (in which case, selective removal and replacement may be appropriate) or is it widespread?

The potential for litigation should be discussed with the building owner prior to conducting the investigation. If litigation is possible, greater documentation is normally required, including documentation related to sampling. However, at the outset of an investigation, the potential for litigation is often unknown. Therefore, it is typically prudent to assume that litigation is possible, and conduct the investigation with this possibility in mind.

Field Investigation

Test cuts should be taken down to the roof deck to determine system composition and condition *(figure 94)*. Two-inch [50-mm] diameter cuts can be taken with a roof coring tool, or small cuts (about 4 inches x 4 inches [100 mm x 100 mm]) can be made. At least one sample should be taken from each roof area where there is a possibility that a different system configuration or type occurs, or if different roof areas are believed to be of different ages. If a re-cover design is contemplated, additional large cuts should be taken as discussed below in Special Reroofing Design Considerations (*page 76*) If the roof is still under warranty, obtain authorization from the warrantor before taking test cuts; otherwise, the warranty may be inadvertently voided.

The field evaluation should include investigating for the presence of asbestos-containing roofing materials *(figure 95)* and the presence of asbestos-containing fireproofing on the underside of the roof deck. Built-up roofs constructed prior to September 1990 often used asbestos felts for the membrane or the base flashings. Prior to that time, asphalt roof cement typically contained asbestos, and some asphalt roof coatings also were asbestos-containing. Asphalt roof cement and coatings are still permitted to

FIGURE 94

During a field investigation prior to reroofing design, a sample was taken down to the deck. Significant dry rot of the tongue-and-groove decking was discovered. Although the underside of the decking was visible from within the building, it did not exhibit signs of deterioration.

FIGURE 95

During a field investigation prior to reroofing design, a sample of the smooth surface BUR was taken to determine if it contained asbestos felts.

contain asbestos, but for the past several years, most of these products have not contained asbestos. Asbestos associated with built-up roofs is typically non-friable. Some early generation thermoplastic single-ply membranes had friable asbestos on the underside of the membrane. Friable insulation is occasionally found around older flue penetrations.

If asphalt is on an existing deck *(figure 96)* and it is desired or required to meet Factory Mutual Research criteria of a maximum of 15 pounds of asphalt per 100 square feet [72 kg/100 m^2], sampling should usually be performed to determine if the existing quantity exceeds this limit. A minimum of three samples per roof area should be taken for laboratory evaluation.

The investigation needs to include a careful assessment of the structural integrity of the deck; for example, has it been degraded by water leakage or condensation? In addition, it is often prudent to assess the roof deck and the supporting structure's resistance to current design seismic, snow or wind loads. The need and scope for this latter assessment are determined by the building's location, age and importance. For example, buildings in areas of low seismic, snow or wind loads would not normally need such an assessment. However, if a building is in a moderate or high seismic, snow or wind area, the architect should determine what building code (or other standard) was used to develop the original design loads and compare them with current load requirements. If the original design loads are significantly lower than current loads, upgrading the deck or supporting structure's load resistance may be prudent even though the upgrade is not mandated by the reroofing provisions in the current code. The building's importance should also be evaluated when considering whether or not to upgrade the structure. For example, if the roof structure of an office building is moderately overstressed when current design snow loads are applied, it may not be

FIGURE 96

Wind blew off this inadequately attached nailer, which caused a large portion of the membrane to blow off. The insulation was attached to the deck with hot asphalt, which was a common attachment method prior to 1984, the year in which Factory Mutual stopped accepting this method. On this deck, only a small amount of asphalt was used, hence removal of the asphalt is not needed to comply with FMR requirements.

necessary to upgrade the structure. But if it is a hospital, school or a critical computer center, it would be prudent to upgrade the structure. For more information about deck uplift resistance, see "Uplift Resistance of Existing Roof Desks: Recommendations for Enhanced Attachment During Reroofing Work" by Thomas Smith.

Nondestructive Testing

Oftentimes the use of one or more of the three nondestructive evaluation (NDE) techniques are extremely helpful in determining the extent of moisture accumulation within the roof system. The following is a brief overview of these techniques:

Electrical capacitance

Electrical capacitance meters create an electrical field below the meter. Moisture in the roof system causes the reading on the meter to increase. Some meters take readings as the meter is rolled across the roof. Other meters only take readings at distinct points. In this case, readings are taken on a grid (usually 10 feet x 10 feet [3 m x 3 m]). Wet areas are more likely to be discovered when a small grid is used, but the smaller the grid, the more costly the survey. Conductive surfaces such as foil-faced membranes and EPDM do not permit the use of capacitance meters.

Infrared thermography

Infrared thermography equipment detects the roof surface temperature. Changes in temperature can indicate wet and dry insulation due to their different heat transfer and storage properties. When using this type of equipment, the entire roof surface is scanned. Infrared surveys can be performed from fixed-wing aircraft or helicopters. Aerial work is economical when a large number of buildings need to be surveyed, such as at a military base or university.

Nuclear hydrogen detection

Nuclear hydrogen detection meters have a radiation source that emits high velocity neutrons when turned on. When the neutrons strike hydrogen atoms, they are counted by the meter. Water molecules and other materials, such as bitumen and concrete, contain hydrogen atoms. Nuclear meters take readings at distinct points, hence this type of survey is also performed on a grid.

Because all three nondestructive techniques indirectly detect moisture, NDE surveys should always be verified by destructive testing, and they should be conducted by a person experienced with the type of technique used. The exception to destructive testing verification is when an existing roof is surveyed before and after new adjacent construction. In this case, destructive testing is only necessary if there is an apparent increase in moisture following the new construction. For further information on NDE, see "A Comparison of Three Different Technologies for Performing Nondestructive Roof Moisture Survey" by Jack Robinson and others.

Historical File

In addition to the field investigation, the evaluation should also include a review of the roof's historical file (which hopefully includes the as-built drawings and specifications, submittals, and previous leakage and repair reports). The investigator should also interview personnel who are familiar with the roof's history to determine, for example, under what conditions the roof leaks.

Checklists

The following resources provide checklists that can be of assistance when evaluating existing roofs:

- *Manual of Roof Inspection, Maintenance, and Emergency Repair for Existing Single-Ply Roofing Systems (SPRI/NRCA 1992)*
- *Manual for Inspection and Maintenance of Built-Up and Modified Bitumen Roof Systems: A Guide for Building Owners (ARMA/NRCA 1996a)*
- *Manual for Inspection and Maintenance of Spray Polyurethane Foam-Based Roof Systems: A Guide for Building Owners (NRCA/SPFD 1998a).*

REPAIR OR REROOF

After the roof has been evaluated, the first decision to be made is whether the roof is a candidate for repair, or if it truly needs to be reroofed. Sometimes spending a few thousand or tens of thousands of dollars, depending upon the roof size, on repair is a prudent course of action. However, it is important to avoid making a significant expenditure on repairs that only provide short-term relief. For example, removal and replacement of a thousand or a few thousand square feet of insulation, depending upon the roof size, that became saturated because of a punctured membrane can be an appropriate solution if the remainder of the roof has several years of remaining service life *(figure 97)*. The distribution of the wet insulation can also influence the repair or reroof decision. For example, if the wet area is contiguous, the removal is more economical than if there are several areas of wet insulation. But if the roof membrane is at the end of its life, it would be prudent to reroof the entire building.

The repair or reroof decision may also depend upon the building owner's intentions. For example, if the owner intends to demolish or sell the building in the near term, repair of a poor roof may be justified, whereas if the owner intends to keep the building for many more years, reroofing would be appropriate.

FIGURE 97

This SPF roof had some wet areas, but it was capable of being saved. A scarifier is being used to plane off the wet foam and prepare it for re-foaming and coating.

If repair is contemplated, it may be wise to

perform laboratory analysis of membrane samples (including field seams, depending upon membrane type). The purpose of this analysis is to obtain additional data that can be used to assess the potential remaining service life. For further information on this topic, see *Condition Assessment of Roofs* (CIB/RILEM 2001a).

Reroofing is necessary when expenditures for repair become excessive, leakage becomes intolerable, the risk of leakage development is unacceptable (in the case of a computer center, for example), or there is catastrophic roof failure (such as wind blow-off or severe hail damage).

TEAR-OFF VERSUS RE-COVER

Advantages and Disadvantages

After it has been decided that the roof will be reroofed rather than repaired, the next decision is whether the existing roof will be torn-off or recovered. There are advantages and disadvantages with both approaches.

Tear-off advantages

- Moisture within the existing roof system is eliminated.
- The entire surface of the roof deck can be checked for structural integrity and attachment.

Tear-off disadvantages

- There is greater potential for water leakage and occupant disruption during reroofing.
- Increased cost related to demolition and debris disposal.
- Increased landfill demands.
- Increased demand for replacement materials such as new insulation to replace the existing.

Re-cover advantages

- The potential for water leakage during reroofing is minimized.
- The R-value of the existing insulation is retained (so the building owner's original investment is not thrown away).
- If the new membrane leaks, the old membrane often prevents the water from reaching the interior of the building. However, this can also be a disadvantage, as minor leakage can go undetected for a considerable time, thus allowing water to migrate between the old and new membranes for a considerable distance.
- In some instances, uplift resistance of the roof system is enhanced.
- This approach is less expensive because demolition labor and landfill disposal costs are avoided.
- Decreased landfill demands.

Re-cover disadvantages

- There is potential for moisture entrapment, which could degrade the roof deck or other roof system components.
- There is limited opportunity for discovering and correcting deteriorated or inadequately attached deck.

Reasons not to Re-cover

In some instances it is unwise to re-cover. These are discussed below:

Building code limitations

The local building code may not permit covering over wet insulation, or it may limit the number of permissible covers. Removal of just the roof membrane may overcome this latter limitation.

Wet insulation

An extensive amount of experimental and field research work has been conducted by ORNL on re-covering over wet roof systems. In some situations, this can be a viable approach. However, until this becomes an industry- and code-accepted practice, re-covering over an existing wet roof system other than for research purposes is not recommended.

Structural limitations

Re-cover is not viable if the roof deck or deck support structure cannot accommodate additional dead load, unless part of the existing system is removed (for example, removal of the loose aggregate on a built-up roof). If the deck is deteriorated or inadequately attached, the existing roof system should be torn off so that the deck can be repaired.

Hurricane-prone regions

In hurricane-prone regions (as defined in *ASCE 7*), it is wise to tear-off rather than re-cover, so that deck integrity and attachment can be verified and strengthened if necessary.

Phenolic roof insulation

If the existing roof contains phenolic insulation, the existing roof system should be torn off to avoid corrosion-induced problems associated with this type of insulation *(figure 98)*.

FIGURE 98

View of a piece of steel deck that has been severely corroded by leachate from wet phenolic roof insulation. The top horizontal flange had separated from the web.

Partial Tear-off

A variation on the re-cover approach is to remove the membrane, but retain the insulation. The advantage of the partial tear-off option is that the entire top surface of the insulation can be observed, which can facilitate detection of wet insulation. However, with this option, some of the advantages of the re-cover approach are sacrificed.

In this variation, after removing wet insulation and infilling with new, a cover board is typically placed over the existing insulation and the new membrane is then installed. If the new membrane is fully adhered, the cover board is typically mechanically attached. If the new membrane is mechanically attached, the membrane fasteners may be sufficient to attach the cover board, depending upon the size of the cover board and the spacing of the fasteners. If the new membrane is ballasted, the cover board is typically loose-laid. A temporary membrane over the tear-off area is typically only required if the new membrane is not installed over the tear-off on the same day that the tear-off is made.

THERMAL UPGRADE

Under certain conditions it is relatively economical to thermally upgrade the existing roof system without actually reroofing. This may be accomplished by changing the reflectivity of the roof surface (addressed in the Energy Efficiency section of the "Design Considerations" chapter) or by adding thermal insulation. Prior to initiating a thermal upgrade, the following evaluations should be performed:

- Verify that the increased thermal resistance will not adversely affect the roof system's fire performance (*see the Fire Performance section of the "Design Considerations" chapter, page 32*).
- Evaluate the existing roof system to determine if the roof has sufficient remaining service life to justify proceeding with the thermal upgrade.
- Estimate the cost of the upgrade and perform life-cycle cost analysis to verify that proceeding with the thermal upgrade is justified.

It may be possible to add thermal insulation below the roof deck. This, however, is rarely a cost-effective solution. If insulation is added below the deck, dew-point calculations should be performed to verify that the dew point occurs above the deck. If the bottom of the deck is at or below the dew point, condensation will occur on the deck. A vapor retarder will need to be installed below the insulation if condensation is a concern. Maintaining vapor-retarder continuity in an under-deck application is difficult. Therefore, in this case it is preferable to omit the retarder and reduce the amount of below-deck insulation so that the dew point occurs above the deck. The other option is to add insulation above the membrane, thereby creating a PMR (discussed in "Membrane Materials").

If the existing roof is a lightly insulated PMR, it too may be upgraded by removing the ballast and installing additional insulation. However, a thermal upgrade is not recommended if the existing insulation is adhered to the membrane because the insulation may float during a rainstorm during or after construction, and, in the process, tear the membrane *(figure 99)*. If the existing insulation is not adhered to the membrane, board floatation does not result in membrane damage.

If a thermal upgrade is performed, and if the HVAC equipment is near the end of its service life, an equipment upgrade should also be considered. With the reduced cooling or heating load, installation of smaller, more energy-efficient equipment would result in additional energy savings.

FIGURE 99

The XEPS in this aggregate-ballasted PMR had intermittently bonded to the flood coat on the BUR. It was not intended for the XEPS to bond to the asphalt, but this is a common occurrence if a polyethylene slip sheet is not inserted between the BUR and XEPS. During a rainstorm, this board floated because the bond was weaker than the buoyancy force. When the board lifted, in some places the flood coat debonded from the top plysheet and in other areas the membrane itself was torn.

TEMPORARY ROOF

On most reroofing projects, a temporary roof is not necessary. Even on tear-off projects, typically a portion of the roof system is torn off and replaced with the new system the same day. However, in some cases it may be cost-effective for the contractor to remove all or a large portion of the existing roof and install a temporary roof. This approach is used when there is an unusual rain threat during construction or to facilitate construction (for example, it can be cheaper to perform tear-off all at one time).

Typically it is not necessary to specify a temporary roof. If the contractor desires to install a temporary roof to minimize cost or risk of water infiltration during construction, the contractor will propose a no-cost change. In this case, the contractor prepares a temporary roofing proposal for the architect's approval. The proposal should include details regarding the materials, temporary system design, and whether or not the temporary roof will remain in place. If it will remain, the proposal should include surface preparation requirements needed to receive the permanent roof and a letter from the manufacturer of the permanent roof membrane stating their acceptance of the temporary roof membrane. The manufacturer's letter should also state that the inclusion of the temporary roof membrane will not adversely affect the fire, wind or FMR rating (if applicable) of the roof system.

In rare instances, the architect may desire a temporary roof. This typically occurs when the roof covers a facility, such as a computer center, that is very sensitive to water leakage. In these cases, rather than rely on the contractor to propose a system, it is typically prudent to specify a conservative temporary roof (including surface preparation requirements prior to installation of the permanent roof).

The NRCA Roofing and Waterproofing Manual includes guide specifications for built-up temporary roofs. However, temporary roofs can also be constructed with modified bitumen and single-ply membranes.

SPECIAL REROOFING DESIGN CONSIDERATIONS

In addition to the design considerations discussed in the previous chapter, the following special considerations are applicable to reroofing design:

- If the existing roof system failed prematurely, the new system design should respond to the problems that caused the premature failure *(figure 100)*. Otherwise, the new system may also fail prematurely.

FIGURE 100

If a roof fails prematurely, it is imperative that the architect determine the reason for the failure before reroofing. The commercial building illustrated here is a case in point. The original roof system was composed of XEPS, preservative treated wood nailers and metal panels. Before application of the original panels, it rained and some water subsequently leaked into the building after the panels were installed. During investigation prior to reroofing, it was discovered that the panels were inadequately attached, so they had to be removed. During reroofing, an air retarder ("housewrap") was placed over the insulation before the panels were installed. Residual moisture within the system would be driven upward during the summer drying months and flow through the air retarder (which is highly vapor permeable). The moisture would then vent out of the system, or if it condensed, it would flow down over the air retarder and exit the roof at the eave. The air retarder also offered rain protection during the reroofing work.

- If portions of the existing roof system are removed, infill materials are needed in the affected areas. Roof insulation is typically needed and vapor-retarder materials are sometimes needed. A membrane patch over the infill area may or may not be needed (this primarily depends upon construction scheduling). In most cases, it is appropriate to specify infill materials to match the existing. However, if the new system will be fully adhered, verify that the attachment of the infill materials complies with the current building code. Also, verify that the infill materials meet current building code fire performance requirements. If an existing component is no longer available, specify a suitable alternative.
- The new system should comply with building code provisions, including those provisions specifically related to reroofing. In particular, be aware of minimum roof slope requirements. Environmental loads (seismic, snow, rain and wind) and dead loads should also be considered, along with fire-resistance and energy-efficiency requirements.
- If compliance with FMG is required or desired, refer to section 2.5 of *Loss Prevention Data Sheet 1-29* for reroofing recommendations. In par-

ticular, be aware of requirements related to steel decks that have asphalt on them. If the roof system was adhered to the deck, several samples should be taken to determine the quantity of asphalt. If the quantity remaining on the deck exceeds the FMG allowable, the excessive asphalt needs to be removed, which is quite difficult and costly. If the quantity does not exceed the FMG allowable, an insulation specifically approved for reroofing by FMR is required.

- Dew-point calculations should be performed to verify that the dew point occurs in an acceptable location. If condensation is a concern, a vapor retarder will need to be added. Even if the existing roof does not have a vapor retarder, the need for one should still be assessed. A change in interior humidity, a different system configuration or different materials could necessitate including a vapor retarder in the new system.
- The occupancy or under-deck conditions (such as the presence of asbestos-containing fireproofing or mechanical or electrical equipment adjacent to the deck) may dictate the method used to attach the new roof system. For example, if exposed fasteners are visually undesirable, or if driving fasteners through the deck could dislodge dust that would be detrimental to occupants or manufacturing operations, or if fasteners could penetrate electrical conduits, then an attachment method that does not rely upon fasteners would be preferable. Or, if fireproofing could be dislodged by use of heavy equipment during application, a ballasted system should be avoided.
- If wood nailers are to be reused (for example, at metal edge flashings or parapets), verify that they are in good condition and adequately attached to meet current wind uplift loads *(figure 93 on page 66).*
- Parapet substrates should be suitably prepared to receive the new roof system. Depending upon the type of new and existing roof systems, it is often prudent to install new preservative-treated or exterior fire-retardant-treated plywood to receive the new parapet flashing.
- If existing penetrations are awkwardly located (for example, a plumbing vent immediately adjacent to a parapet) it is prudent to relocate the penetrations, rather than try to execute details that are unlikely to provide long-term service *(figure 101).* Alternatively a SPF system could be specified, as this system readily lends itself to awkwardly located penetrations and to roofs scattered with piping *(figures 102 and 103).* If HVAC equipment is on sleepers or stands that are too low, the

FIGURE 101

This is an example of a penetration that should be revised when reroofing. The conduit should be raised and supported on curbs. The junction box should also be raised.

equipment should be raised. If door thresholds at doors leading out onto the roof are too low, they should be raised. If there are weep holes, the new system should not block the weeps (this may require modifying the wall to raise the weeps).

- Existing mechanical and electrical equipment and skylights that have been abandoned should be removed. If the equipment or skylights are curb-mounted, the curb should be removed and the deck opening filled in. Retaining the curb and providing a cover over the curb opening may be a less expensive option, but is not as desirable because a membrane penetration still occurs. If an abandoned curb is retained, a sheet metal covering over the curb can be used if the opening is not too large. Insulation (and vapor retarder when needed) should be provided under the covering. Decking material is used to span larger curb openings. In this case, the decking is normally covered with the same roof system used for the field of the roof.
- If a more energy-efficient roof system is designed, consider upgrading the HVAC equipment as previously discussed in the Thermal Upgrade section (*page 74*).
- If the roof is on a historical building, special consideration may be necessary to comply with preservation goals, while delivering a roof system that offers higher performance. For example, use self-adhering modified bitumen in lieu of organic felt for underlayment beneath flat-seamed soldered architectural metal panels. This approach provides a high performance underlayment for secondary water infiltration protection while offering a roof covering that replicates the original metal roof.

FIGURE 102

A large number of pipes and conduits penetrated this wall and ran parallel to it. This situation was relatively easy to flash with SPF; it would not have been with other membrane materials.

FIGURE 103

This roof had a large amount of piping on elevated curbs and numerous pieces of HVAC equipment. It was relatively easy to roof with SPF; it would not have been with many other membrane materials.

Re-cover Design

In addition to the above, the following special considerations are applicable to re-cover design:

- NDE should always be performed (except for those systems where

NDE is not applicable) during the design stage. Areas of wet insulation should be identified and specified for removal.

- Several large test cuts (2 feet x 2 feet [600 mm x 600 mm] minimum) should be taken to assess deck integrity and attachment during the design stage. The number of cuts will depend on several factors, including the deck type, roof size, leakage history and extent of wet insulation. Where possible the underside of the deck should also be evaluated.
- There are two basic ways to attach the new system. One approach is to keep it divorced from the existing roof. The other approach is to adhere the new system directly to the existing membrane.

  The divorced approach can be accomplished in several ways: by mechanically attaching a layer of insulation (or loose-laying it in the case of a ballasted system) between the old and new membranes; by using a fleeced-backed membrane that is ballasted, mechanically attached, or in a PMR configuration; or by installing a metal panel system over a new support structure above the existing roof. With a divorced system, the new system is not dependent upon the existing system for uplift resistance.

  The adhered approach is commonly done with sprayed polyurethane foam. With this technique, both the existing and the new systems need to possess adequate uplift resistance. When such an approach is contemplated, extensive field uplift resistance testing should typically be performed during the design stage. If the existing system is inadequately attached, it may be possible to mechanically attach the existing membrane and then adhere the new system to it.

  When a new system is adhered to the existing, attention needs to be given to cleaning the existing membrane. In most cases it is also prudent to specify priming of the existing membrane.
- SPRI recommends considering slitting or removing the existing roof membrane if the existing membrane is not needed as a vapor retarder. The advantage of slitting or removing the membrane is that water will not accumulate between the old and new membranes if the new roof leaks. Other advantages of membrane removal were previously discussed. SPRI, however, advises that the disadvantage of slitting the membrane is that moisture within the existing roof system may migrate up and accumulate between the old and new membranes.

  As discussed above in Tear-off Versus Re-cover (*page 72*), there are advantages and disadvantages with slitting or removing the membrane. In many instances it is preferable not to slit or remove the existing membrane, but this issue should be evaluated for each re-cover project.

REROOFING ABOVE OR ADJACENT TO AN EXISTING ROOF

Sometimes a building will have one or more areas that are in need of reroofing, but other roof areas may not need to be reroofed. Reroofing above or adjacent to an existing roof that is not included in the reroofing project has the very great potential of causing harm to the roof that is to remain. If the roof that is not scheduled for reroofing has limited service life remaining, it is prudent to specify that it too be reroofed. If the roof has ample life remaining, the architect should consider the recommendations discussed below. Even when following these recommendations, however, the architect will find that saving an existing roof is challenging. In many cases it may be best to avoid the NDE and protective measures and include the roof area in the reroofing project.

- Have a non-destructive evaluation (NDE) performed shortly prior to construction. The purpose is to determine if there is wet roof insulation.
- Specify protection requirements. These will vary according to the type of existing roof and expected loading during construction. In many instances, installing plywood or OSB over a cushion (such as an inch [25 mm] of MEPS) will suffice.
- If the roof is under warranty, contact the warrantor to determine if they have protection recommendations or requirements. Also determine if the warrantor desires to inspect the roof after construction is complete.
- During a pre-construction conference, discuss the importance of taking care of the roof, observe the roof with the contractor and agree on its condition before construction. Advise the contractor of the results of the NDE and advise that a NDE will be performed after completion of the work.
- After job completion, have another NDE performed and observe the roof. If the NDE shows new areas of wet insulation or if damage is observed, corrective work should be undertaken. Also if the roof is under warranty and the warrantor desires to inspect the roof after construction is complete, verify the inspection occurred and that the warrantor agrees to keep the warranty in force.
- Advise the building owner to have another NDE performed two years after job completion, as damage may not immediately manifest itself.

SPECIAL ASPECTS OF SPECIFICATIONS AND DRAWINGS

In addition to information routinely included in specifications and drawings for new roofing projects, the specifications and drawings for reroofing projects should address the following:

- The existing system (including component materials and approximate thickness of each type of insulation) should be identified so that the contractor can adequately estimate the job.
- The documents should indicate if asbestos-containing roofing, fireproofing or pipe insulation materials occur *(figure 104)*. If non-friable materials are to be removed, it is preferable for the work to be performed by a roofing contractor. If friable materials are to be

FIGURE 104

During the design of a reroofing project, it is important to check for asbestos-containing materials. In this example, the upper pipe, which was supported from the roof structure, was covered with asbestos-containing insulation. The insulation jacket, which was torn at the pipe hangar, should be repaired, or the insulation should be removed prior to reroofing.

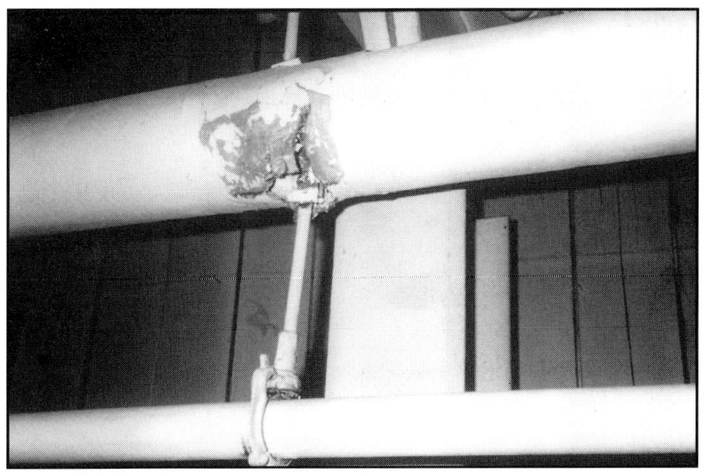

removed, they need to be removed by a separate abatement contractor, unless the roofing contractor is qualified to remove friable asbestos. If an abatement contractor performs removal, specify coordination between that contractor and the roofing contractor so that the building is not susceptible to water infiltration during or after the abatement work is completed. The documents should also provide guidance if asbestos-containing materials are unexpectedly found.

- If fireproofing on the underside of the deck is poorly attached and is likely to be spalled off by the reroofing work, the documents should address remedial fireproofing work.
- Specify if excessive asphalt needs to be removed from a steel deck. Furnish the sampling test results so that the contractor can determine how much asphalt needs to be removed. If compliance with FMR criteria is desired or required, specify an insulation specifically approved by FMG for reroofing. Asphalt removal is very expensive.
- Specify that the contractor is to take photographs or a videotape of existing conditions, including interior and exterior building surfaces, sidewalks, paving and landscaping that might be misconstrued as damage caused by the reroofing work. Require submittal of the documentation prior to beginning work.
- Specify protection of persons from injury, and protection of the building, sidewalks, paving and landscaping from damage or soiling.
- Specify that the contractor coordinate work activities on a daily basis with the owner in those situations where the owner desires to place protective dust or water leakage covers over sensitive equipment or furnishings or where it is desired to evacuate people from underneath the work area.
- If deck removal is required, special requirements related to occupant protection and protection from leakage should be specified. Occupant protection typically can be achieved by evacuating occupants from

beneath the work area. But in rare instances, very special protective measures may need to be specified. For example, where occupant evacuation is not possible, a protective platform can be constructed below the roof deck. If the building is located in an area that experiences very frequent rainstorms, a weather-tight work enclosure may need to be positioned over the work area as illustrated in *figures 105 and 106.*

FIGURE 105

This building, which is in an area that experiences rain and moderate winds nearly every day, had to have its deck replaced while remaining open for business. To provide a weather-tight work enclosure, the contractor constructed a wood box, which was lifted into place *(see figure 106).*

FIGURE 106

View of the temporary enclosure after it was anchored and flashed to the roof. After reroofing work was completed in one area, the box was moved to another area on the roof.

- If HVAC equipment will need to be shut down during roof application, specify that the shutdown be coordinated with the building owner. If air intakes are in the work area and the work is likely to result in objectionable odors or fumes, specify that the intakes are to be shut down and covered while work is underway in the vicinity of the intakes.
- If fire detection devices need to be shut down (for example, duct or plenum smoke detectors) because of concern about activation due to dust or fumes, specify coordination of the shutdown procedures with the building owner. Specify if special measures need to be taken by the contractor during the deactivation of portions of the fire detection system, such as implementation of a fire watch or working only during special hours or days.
- Specify that each roof area has suitable roof drainage at the end of each day. For example, if roof drains are blocked during the day to keep out

debris, they should be unblocked before the roofing crew leaves for the day. Also, if the existing drainage pattern is blocked by demolition or partial installation of the new system, some method for eliminating ponding needs to be incorporated so that nighttime rainfall does not result in roof collapse.

- If the new components, such as insulation or membrane, are mechanically attached, specify that fastener pull-out testing be performed in accordance with *Standard Field Test Procedure for Determining the Withdrawal Resistance of Roofing Fastener,* also known as *ANSI/SPRI FX-1*, before installing the new system. Specify testing by the roofing contractor or by an independent testing agency.
  During design, some limited pull-out testing should be performed to provide data for design purposes.
- If a temporary roof is required or reroofing occurs above or adjacent to an existing roof, refer to the previously discussed Temporary Roof and Reroofing Above or Adjacent to an Existing Roof sections (*pages 75 and 79*) for additional specification and drawing requirements.
- Peer review should be considered on reroofing projects that are very complex, very costly or in those cases where the architect has limited expertise and experience with reroofing design.

SPECIAL ASPECTS OF CONSTRUCTION CONTRACT ADMINISTRATION

In addition to activities routinely associated with construction contract administration related to new roofing projects, the following are applicable to reroofing projects:

- During the pre-roofing conference, the roofing contractor's foreman should be advised under what conditions the architect should be consulted concerning unforeseen conditions. For example, deck integrity parameters should be established. Generally the architect would not need to be consulted if superficial surface corrosion is found. But if advanced corrosion is discovered, the need for corrective action should typically not be determined by the contractor.
- The architect (or other entity engaged to provide construction contract administration services) needs to be able to quickly respond to unforeseen conditions in order to avoid idling the roofing crew.
- The architect should check to see if appropriate measures are being taken with respect to occupant and leakage protection. Verify, for example, that the contractor is notifying the building owner about which areas are being worked on each day. Verify that excessively large areas of membrane are not being torn off. (In other words, the tear-off should be small enough for the contractor to make the area watertight in the event of rain.) And verify that air intakes are shut down when so needed.
- If the architect observes that satisfactory measures are not being taken to protect the building or grounds, the contractor should be advised.

SOURCES OF ADDITIONAL INFORMATION

The following manuals and books have further information on reroofing:

- *The NRCA Reroofing Manual.* This very comprehensive document should be in the office of every architect who is involved in reroofing design. It is included in *The NRCA Roofing and Waterproofing Manual.*
- *The Manual of Low-Slope Roof Systems* by C.W. Griffin and R.L. Fricklas.
- *Proceedings of the Low-Slope Reroofing Workshop* by ORNL.
- *Metal Roofing Systems Design Manual* by MBMA.

SUSTAINABLE DESIGN CONSIDERATIONS

This chapter provides an overview of sustainable roof principles as well as some design recommendations. Sustainable roofing is a subset of sustainable construction, which has been a developing movement. One milestone of the movement was the 1992 United Nations Conference on the Environment and Development in Rio de Janeiro. The Conference adopted "Agenda 21," an environmental and development program for the twenty-first century. The principles of "Agenda 21" are defined in *The Rio Declaration on Environment and Development.* The *Declaration* is attached to "Sustainable Roofing: A Paradigm of Challenges and Opportunities" (Smith 1997). Two particularly noteworthy ideas from the *Rio Declaration* are:

- "Development today must not undermine the development and environmental needs of present and future generations."
- "In order to achieve sustainable development, environmental protection shall constitute an integral part of the development process, and cannot be considered in isolation from it."

Other terms used in lieu of *sustainable construction* are *environmentally friendly construction*, *green construction* and *green buildings*. In the roofing industry, however, the term *green roofs* refers to a specific type of roof system, as discussed in the "Design Considerations" chapter. A green roof can be designed and constructed in a manner that would be considered sustainable. But a green roof can also be designed with little regard to thermal efficiency, and it can be designed or constructed in a manner that can lead to premature failure. Hence, a green roof should not automatically be thought of as a sustainable roof.

As with the design considerations discussed in an earlier chapter, sustainable design considerations influence both the selection of the roof system and how the system is developed and detailed.

ELEMENTS OF SUSTAINABLE ROOFING

Sustainable roofing issues are complex and involve interrelationships that are not well understood at present. The foreword to ORNL's *Proceedings of the Sustainable Low-Slope Roofing Workshop* defined a sustainable roof "as a low-slope roofing system that is designed, constructed, maintained, rehabilitated, or demolished with an emphasis throughout its life-cycle on the efficient use of natural resources and the maintenance of the global environment." In keeping with this definition, the full implementation of the sustainable roofing concept requires consideration of the following: raw materials extraction and processing, material production, material packaging, material transportation, roof system design, installation, maintenance, repair, recover or tear-off at the end of the roof's life, and reuse, recycling or disposal of the tear-off materials. These various interrelated factors can be grouped into six primary elements: materials, design, installation, roof service life, thermal performance and material disposition.

Materials

Material selection for sustainable roofs is based on the assessment of four main factors: embodied energy, environmental consequences, durability and material disposition when the service life of the roof is over. The first two factors are discussed here. The last two factors are discussed later in the chapter sections Roof Service Life and Material Disposition.

The embodied energy of a product is a measurement of all of the energy expended for acquiring the raw materials, transporting the raw materials to the manufacturing facility, processing them, transporting the finished product to the construction site and the inherent energy in the product. For roof membrane materials, embodied energy is reported as the number of Btus per square foot [MJ/m^2]. As of the year 2001, embodied energy data are not available for most roof system products.

In addition to embodied energy, it is also important to know other environmental consequences associated with a material. For example, until the early 1990s, some plastic foam insulations were blown with chlorofluorocarbon (CFC). When released to the atmosphere, CFC causes stratospheric ozone depletion, the loss of which results in increased harmful ultraviolet radiation striking the earth's surface.

Traditionally, material selection has been primarily based on assessing the ability of the materials to perform and on their costs. In sustainable construction, materials are selected with respect to their cradle-to-gradle profile (in other words, from raw materials acquisition to their eventual reuse, recycle or disposal). The term *gradle* is reflective of the fact that at the end of a roof's life, some of the materials will probably require disposal (they will be placed in their grave) and some of the materials may be reused or recycled (their life will begin again in a new cradle). The cradle-gradle model for roofing is illustrated in *figure 107*.

Material selection based on cradle-to-gradle analysis is exponentially more difficult than the current material selection process.

FIGURE 107
Cradle-to-Gradle Model for Roofs

1. **Materials Production**
   - Raw Materials (including virgin and recycled): extraction, processing, transportation
   - Material Production
   - Material Packaging
2. **Transportation of Materials to Construction Site**
3. **Design of the Roof System**
4. **Installation of the Roof System**
   - Generation of Construction Waste: recycle or disposal
5. **Roof System Service Life**
   - Periodic Maintenance
   - Repairs (as needed)
6. **Reroofing**
   - Re-cover or Tear-off
7. **Tear-off Debris**
   - Reuse, Recycle or Disposal

Design

Compared with material production or roof application, the design phase of a project typically consumes the smallest amount of energy.

Yet the materials and system selection and detailing decisions made by the architect have significant positive or negative environmental consequences. Design decisions directly affect the thermal performance of the roof and the potential for material reuse/recycling. These decisions also affect the roof service life, although service life is also influenced by other factors.

Because of the lack of design guidelines on sustainable roofing and lack of various types of information (such as embodied energy and credible data on service life), it currently is difficult to design a roof that is responsive to all elements of sustainable construction. Unfortunately, the little design information that is available may be unreliable. Because the concept of sustainable roofs and the term itself are relatively new to the roofing industry, many roof consultants, manufacturers and contractors either misunderstand or only partially understand the comprehensiveness of sustainable roofs. To some people, a sustainable roof means a component of the roof system is made from recycled materials. However, recycled content is only a minute aspect of sustainable roofs.

Because guidelines or protocols for assessing divergent issues presently do not exist, it is sometimes difficult to make informed decisions. For example, perlite insulation has recycled newsprint, whereas polyisocyanurate insulation does not incorporate post-consumer waste. At first glance, it might seem reasonable to specify perlite for the thermal insulation. But if an R-value of 20 is desired, 9.6 inches [244 mm] of perlite would be required versus 3.6 inches [91 mm] of polyisocyanurate. While use of perlite for a cover board is often appropriate, in most cases it would not be prudent to use perlite to achieve a large amount of thermal resistance.

Another example is an aggregate-ballasted single-ply. The ballast can be reused in the future, and it could be possible to easily remove and recycle loose-laid MEPS. But if the building is a hospital in a hurricane-prone region, this would be a poor system choice. Also, if the roof has a lot of foot traffic to maintain HVAC equipment, and if the system does not incorporate a stone-protection mat, the membrane could become punctured and no longer serviceable within a few years.

Recommendations for architects desiring to emphasize sustainable roof design are provided at the end of this chapter.

Installation

Two aspects of installation are significant to sustainable roofing. First and most important, the quality of the application can be a key determinant of the roof's service life. Second, construction waste generated as a result of application can be handled in a sustainable or non-sustainable manner. In the United States, construction waste is typically placed in a landfill. At least in densely populated areas, however, there are opportunities to recycle various types of waste.

Roof Service Life

Roof service life is a function of roof durability, the degree of maintenance and repair received by the roof, level of abuse *(figures 108 and 109)* and severe weather events. As previously discussed in the "Design Considerations" chapter, durability refers to the capability of the roofing system to remain serviceable (functional) over a minimum specified time. More durable roof systems have less negative environmental impact because of the reduced energy consumption and waste generation associated with less frequent reroofing.

FIGURE 108

This roof is being abused by indiscriminate placement of the HVAC equipment screens directly on the roof. Such disregard can easily shorten the life of a roof.

FIGURE 109

The roof of this manufacturing facility has become a work platform: Pallets and heavy pieces of metal are resting directly on the modified bitumen membrane. Although this type of membrane is relatively abuse-resistant, an ink pen in the foreground marks the location of a recent membrane puncture.

Thermal Performance

Except for high-rise buildings, the thermal performance of the roof system typically plays a major role in the amount of energy the building consumes for heating or cooling. Considering that buildings consume 30 to 40 percent of the total energy in the United States (excluding embodied energy), the potential energy savings offered by roof systems is tremendous. Buildings with thermally efficient roofs save money for building owners and decrease the demand on fossil fuels, the burning of which degrades the environment.

Thermal efficiency can be achieved with a roof assembly that has a low U-value (high R-value), with a roof surface that is highly reflective (as discussed in the "Design Considerations" chapter) or a combination of both.

Material Disposition

At the end of a sustainable roof's life, much of the roof system should be capable of being reused or recycled; as little material should go to a landfill as possible. The preferred option is to reuse some of the old system in the reroofing project. Ballast, for example, is a component that is easily reused. Extruded expanded polystyrene from PMRs can also be easily reused, although the moisture content of a few of the

FIGURE 110

This is an EPDM in a PMR configuration. It should provide a very long service life. At the end of the membrane's life, the XEPS insulation boards can be easily removed. Most of the boards should be in good enough condition to be reinstalled over a new membrane. Those that are too wet to reuse can be easily recycled (assuming a recycling facility is in the vicinity).

boards may be too great to allow reuse. MEPS is an example of a product that can be recycled. However, if it is contaminated with other material (such as a cover board adhered to the MEPS with asphalt), recycling is currently uneconomical.

The key to a successful reuse/recycle scenario lies at the beginning of the project when the materials and system design are selected *(figure 110)*. Decisions made 20 to 40 years in the past will have a profound impact on the amount of the old roof system that needs to be hauled to a landfill. For example, MEPS from a ballasted or mechanically attached single-ply system is easy to recycle. However, MEPS with an asphalt-adhered cover board from a modified bitumen system is not.

DESIGN RECOMMENDATIONS

In 2000, an international roofing committee working under the auspices of CIB/RILEM published a set of common principles for sustainable roofing (CIB/RILEM 2001b). The committee identified 21 tenets *(figure 111)*, which they grouped under three subheadings: minimize the burden on the environment, conserve energy and extend roof life span. Each tenet is described by a single sentence so that designers can quickly determine the major issues. Accompanying the list of tenets is a description that more thoroughly explains each tenet. Architects interested in emphasizing sustainable roof design are encouraged to download the full publication, *Towards Sustainable Roofing*, from www.cibworld.nl.

CIB/RILEM Tenets

Some elements of sustainable roofing require the development of new information and technologies. However, several of the CIB/RILEM tenets can be implemented now. The most important of these are:

- Explore design solutions that will facilitate reuse or recycling at the end of the roof's life (*see previous discussion under Material Disposition for examples*), and develop details that facilitate future reroofing. For example, design the height of parapets and counterflashing receivers so that the roof could be re-covered in the future without infringing upon the desired distance from the membrane to the bottom of the coping or counterflashing.
- Consider self-drying roof solutions to help eliminate water that leaks into the roof (*see Self-drying Roofs in "Design Considerations" chapter, page 51*).
- Design thermally efficient roof systems, with consideration given to both R-value and reflectivity, as appropriate for the climate where the building is sited (*see the Energy Efficiency, Insulation Influences and Vapor Retarders sections in "Design Considerations," beginning on page 46*).

MINIMISE THE BURDEN ON THE ENVIRONMENT

1. Use products made from raw materials whose extraction is least damaging to the environment.
2. Adopt systems and working practices that minimise wastage.
3. Avoid products that result in hazardous waste.
4. Recognise regional climatic and geographical factors.
5. Where logical, use products that could be reused or recycled.
6. Promote the use of "green roofs" supporting vegetation, especially on city centre roofs.
7. Consider roof designs that ease the sorting and salvage of materials at the end of the life of the roof.

CONSERVE ENERGY

8. Optimise the real thermal performance, recognizing that thermal insulation can greatly reduce heating or cooling costs over the lifetime of a building.
9. Keep insulation dry, to maintain thermal performance and durability of the roof.
10. Use local labor, materials and services wherever practical to reduce transportation.
11. Recognise that embodied energy values are a useful measure for comparing alternative constructions.
12. Consider the roof surface color and texture with regard to climate and the effect on energy and roof system performance.

EXTEND ROOF LIFE SPAN

13. Employ designers, suppliers, contractors and tradespeople who are adequately trained and have appropriate skills.
14. Adopt a responsible approach to design, recognising the value of the robust and durable roof.
15. Recognise the importance of a properly supporting structure.
16. Provide effective drainage to avoid ponding.
17. Minimise the number of penetrations through the roof.
18. Ensure that high maintenance items are accessible for repair or replacement.
19. Monitor roofing works in progress and take corrective action as necessary.
20. Control access onto completed roofs to reduce puncture and other damage, providing defined walkways and temporary protection.
21. Adopt preventative maintenance, with periodic inspections and timely repairs.

FIGURE 111

Tenets of Sustainable Roofing

These tenets are applicable to membrane roofing systems on permanent buildings. As our understanding develops of how roofs perform and how they affect the global environment, then the tenets will also evolve. *Reprinted by permission from CIB/RILEM's* Towards Sustainable Roofing.

- Design for enhanced durability and reliability (*see the Durability and Reliability sections in the "Design Considerations" chapter, page 64*). Reconsider your expectations for roof service life by increasing your current mindset by 50 percent. For example, if you currently expect a service life of 20 years, increase your expectation to 30 years and design and specify accordingly.

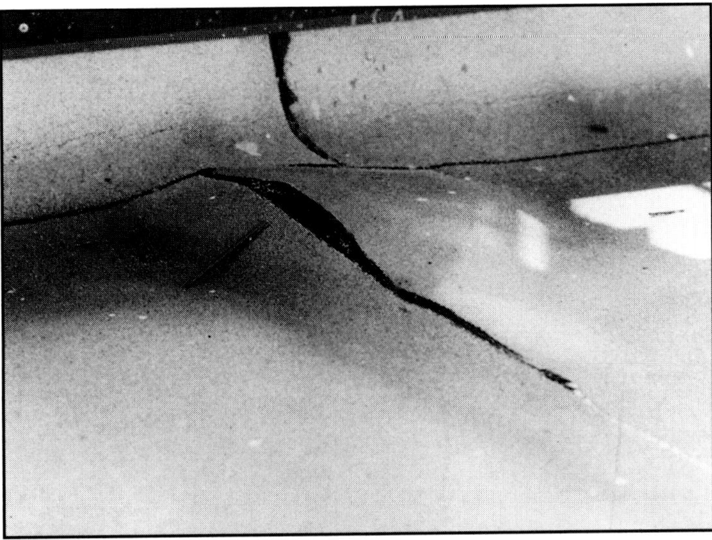

FIGURE 112

If a roof is to achieve a long service life, many conditions need to occur. The building owner, for example, must ensure that the roof is periodically inspected and maintained, and that repairs are executed in a timely manner. This modified bitumen membrane has a very large blister. Near the top of the blister, most of the lap has been pulled apart. Although blisters should usually be left alone, this is an example of a condition that should be repaired prior to seam failure and water leakage.

To achieve significantly increased service life, excellent quality needs to be sought in the roof design and detailing, the materials, application, semi-annual inspections, maintenance, and repair (*figure 112*). Enhancements need to be incorporated into the system (for example, the addition of a stone-protection mat in an aggregate-ballasted single-ply system) and into the details (*see Flashings and Penetrations in the "Design Considerations" chapter, page 51*). The particular system, system enhancements and detail enhancements that will be needed will depend upon several factors, including the climate where the building is located. These other factors and the ways to respond to them during design, application and upon job completion are discussed throughout this monograph.

Additional Suggestions

Architects involved in reroofing design should also consider emphasizing sustainable design concepts in these projects because roof systems are periodically replaced, over a few decades a high percentage of the existing roof inventory can be converted from non-sustainable to sustainable construction.

For further information on sustainable design considerations, refer to ORNL's *Proceedings of the Sustainable Low-Slope Roofing Workshop,* NCARB's *Sustainable Design* and "Designing for Future Reroofing: Recommendations for Enhanced Sustainability" by Thomas Smith.

SYSTEM SELECTION CRITERIA

Not long ago, roof system selection was simple. Built-up roofs were specified for almost all low-slope roofs. Selection issues were primarily limited to the type of bitumen to use (asphalt or coal tar), the ply reinforcement material, the type of membrane surfacing and the type of insulation (with the choices being very limited). This began to change in the 1970s with the emergence of plastic foam insulations, the introduction of modified bitumen and thermoplastic membranes from Europe, and the growth of thermoset membranes, sprayed polyurethane and low-slope metal roofing. By the 1980s, the diversity of systems made the selection of the most appropriate roof system for a specific project a daunting task. The selection process is difficult because of the large number of factors, many of which are unrelated or conflict with one another, and the lack of key data (such as realistic design service life). A computer-integrated knowledge-based (also known as *expert*) system would greatly benefit the selection process, but such a system has not been developed for roofs.

For most roofs, several different types of systems could serve quite well. But some roofs have unique characteristics that lend themselves to perhaps only a few systems. In order to select the most appropriate system for a project, ideally the architect should have at least a general understanding of the membrane options described in the "Membrane Materials" chapter, as well as a general understanding of built-up roofing. Unless the architect is aware of all the different options available, he or she may overlook the most appropriate system for a particular project.

If the architect lacks a general understanding of low-slope system options (particularly if the project is complex, unusual, very expensive or is a reroofing project), the architect should seek system selection input from a professional roofing contractor or professional roof consultant who is knowledgeable about all system options.

In the context of this chapter, system selection refers to selection from the primary system types discussed earlier in the "Membrane Materials" chapter (modified bitumen, single-ply, sprayed polyurethane foam, liquid-applied, and metal panels) and BUR. It also refers to selection of membrane materials within system types (type of bitumen or modified bitumen, type of single-ply membrane, type of surfacing on an SPF, or type of metal panel profile) and, where applicable, the attachment configuration (fully adhered, ballasted, mechanically attached, PMR, or loose-laid air-pressure equalized).

In this chapter, system selection criteria are discussed first. The selection of the manufacturer is addressed thereafter. Finally, a matrix is presented that can be used to assist in selecting the system and manufacturer.

SYSTEM SELECTION

With a general understanding of the available system options, consideration of the following technical and non-technical criteria can lead to the selection of the most appropriate system and details for a project:

- system demise
- contractor familiarity and availability
- maintenance intensity

- owner familiarity
- technical considerations
- cost
- warranty
- implications of sustainable roof design.

The first seven criteria have been listed in order of importance, per the author's opinion. If the architect is interested in emphasizing sustainable design principles, sustainable design criteria would move up on the list (perhaps to third place). It is critical that the selected system sufficiently satisfies all of the criteria.

System Demise

The first step in the selection process is to determine what will likely cause the death of the roof system. Is the project, for example, located in an area that experiences frequent and damaging hailstorms? Will the roof experience constant or extremely strong winds? Does the roof have numerous HVAC units, the service of which will generate perpetual abusive foot traffic? Will the roof be exposed to intense solar radiation throughout most of its life? In some cases, one factor will likely cause system death. In other cases, perhaps two or three factors may be nearly equally as likely to end the roof's life.

After identifying the likely causes of death, it is incumbent upon the architect to select a system with characteristics that can combat these destructive forces.

Contractor Familiarity and Availability

Good application is crucial to the long-term success of a roof. During the system selection process, therefore, the architect should consider the following:

Contractor familiarity with proposed system

Are contractors in the vicinity of the project site familiar with the systems being considered? If not, either a system should be selected that the contractors are familiar with, or a contractor should be brought in from outside of the project vicinity (this last option has downsides, as noted below). It is important to avoid having a contractor install a system that he or she is not extremely familiar with.

Contractor proximity to project site

It is preferable to have a contractor that has an office relatively close to the project site. The contractor will be familiar with the weather conditions, will be available to provide periodic inspection and maintenance if the owner desires to contract for this service *(see "Key Elements of Specifications and Drawings" chapter, page 109)*, and will be readily available to provide repairs if needed.

In some instances, because of size or complexity, the project demands may be beyond the capabilities of the local contractors. In this case, it may be prudent to have the work executed by a contractor outside of the project locale.

Maintenance Intensity

Maintenance is discussed in the "Design Considerations" chapter. Maintenance intensity is listed as a separate system selection criterion because it is a factor that architects frequently overlook when selecting a roof system.

It is important for the architect to try to determine how committed the building owner will be to having periodic inspections and maintenance performed. If the owner is likely to devote little or no attention to the roof, a roof system should be selected that has limited maintenance demands.

Owner Familiarity

If the building owner has only one or two buildings, owner familiarity with different types of roof systems generally is not an issue, although it is still wise to determine if the owner has a preference (or dislike) for a specific system, and why the owner has such a preference. Be cautious of discarding a system from consideration simply because the owner has had problems with it. The problems may have been caused by poor design, materials or application. It is important to try to avoid discarding an appropriate system from consideration because of poor work quality, unless it is believed that poor quality of work will be experienced again.

If the owner has multiple buildings, as in the case of a school district, the architect should request information on the type of systems, number of roofs within each system type, and the experience that the owner has had with the various types. If a specific system has been a good performer, it is probably best to use that system on the upcoming project, unless the new project has unique characteristics that another system would be better able to accommodate. If the owner has periodic inspection, maintenance and minor repairs performed by in-house maintenance personnel, one advantage of keeping with the same type of system is that they will not have to become familiar with another system type. There have been many instances where owner maintenance personnel have inappropriately used BUR repair techniques on a single-ply membrane because they did not know better.

Technical Considerations

Most of the topics discussed in the "Design Considerations" chapter are technical in nature. Many of those considerations strongly influence the system selection. The considerations that influence system selection vary from job to job, depending upon the project location and requirements.

When selecting a system, it is important for the architect to determine whether the proposed system should more than just meet the minimum requirements. For example, if external fire resistance is particularly important for a project, then specifying a system that requires a field-applied coating to achieve the rating is probably not the best choice, as the fire resistance declines as the coating weathers away unless the owner is diligent about re-coating *(figure 113)*. A better choice would be a system that achieves a Class A rating without a coating. If enhanced protection is desired, a metal roof or paver-ballasted system could be the best choice.

FIGURE 113

This aluminum-pigmented asphalt roof coating is weathering away as expected. Before specifying a coating, the architect should be confident that the building owner will have the roof periodically re-coated. Otherwise, a system that does not require a coating should be specified.

Cost

Many architects and building owners select a roof system primarily on initial cost. Obviously cost is an important element of a project, but the ramifications should be understood when cost is a governing factor in system selection. If a less expensive system is selected, invariably something suffers in comparison with the systems that fell from consideration because of the greater cost. The cheaper system generally will not have the reliability or durability of other systems, it may be more maintenance intensive or it may not be as energy efficient. Over the life of the roof, the system with the lowest initial cost often is more expensive than other options that were discarded because of their higher initial cost. The table in *figure 114* provides an approximation of the cost differential of the various types of low-slope roof systems.

In evaluating cost, it is important to look at the life-cycle cost (LCC). In addition to the initial construction cost, LCC includes energy consumption (for building heating and cooling), maintenance, repairs, length of design service life, and disposal at the end of the roof's life. Of these factors, the most difficult to assess is the design service life.

The service life can have a dramatic impact on the LCC analysis. For example, if a 40-year service life is assumed, but the roof fails after 15 years, the owner's true roofing costs will be much higher than calculated. Lack of good data on design service life is often a significant limitation to developing a reliable LCC. For example, it is difficult to have confidence in a manufacturer's claim of a 30-year life for products that have been in the marketplace for only a few years. Accelerated aging testing is of limited help, as it has not progressed to the point where credible estimates of service life prediction can be made. The selection of a predicted service life should be conservative. For most low-slope systems, use of a service life in excess of 20 years should only be done with caution, evaluation and justification. If the roof lasts five or 20 years longer than estimated, the building owner will be less concerned than if it fails in half of the estimated time.

For most projects, the costs associated with eventual tear-off and disposal are seldom considered. Because some systems are inherently more difficult to tear-off than others, LCC analysis should consider this issue. Also, it may be possible to salvage or reuse some of the system components. For example, with a ballasted single-ply, it would be reasonable to assume that the ballast could be reused on the replacement roof.

FIGURE 114
Approximate Cost Differential of Low-Slope Roof Systems

SYSTEM TYPE	COST IN COMPARISON TO AGGREGATE-SURFACED BUR[1]
BUR, aggregate-surfaced[2]	1.0
Modified bitumen[3]	1.1
EPDM, aggregate-ballasted[4]	0.73
EPDM, fully adhered[5]	0.94
PVC, mechanically attached[6]	0.86
SPF, coated surface and granules[7]	1.07
Structural metal panel[8]	2.83

NOTES:

1. Caution: For a specific roof, a variety of factors can significantly alter the cost differential shown. These cost comparisons assume the following:
   - New building located in a suburban area near Denver, CO.
   - Building is two stories high, with a roof area of 50,000 square feet [4645 m^2].
   - Steel deck.
   - Target for insulation R-value: 18.
   - Very few roof penetrations.
   - Perimeter edge condition: 12-inch-high [305 mm] parapet.
   - Roof slope: ¼:12 [2 percent], achieved by sloping the deck.
2. Aggregate-surfaced BUR system: Four-plies of fiberglass felt set in hot asphalt, over a ½ inch [13 mm] cover board set in hot asphalt, over two layers of 1½ inch [38 mm] polyisocyanurate insulation (3 inch [75 mm] total thickness) mechanically attached to the deck.
3. Modified bitumen system: Granule-surfaced modified bitumen cap sheet set in cold adhesive, over a modified bitumen base sheet set in cold adhesive over a ½ inch [13 mm] cover board mechanically attached through two loose-laid layers of 1½ inch [38 mm] polyisocyanurate insulation (3 inch [75 mm] total thickness) over the deck.
4. Ballasted EPDM system: Aggregate over a stone-protection mat, over a loose-laid 45 mil [1.15 mm] non-reinforced membrane, over two loose-laid layers of 1½ inch [38 mm] polyisocyanurate insulation (3 inch [75 mm] total thickness) over the deck.
5. Fully adhered EPDM system: 60 mil [1.5 mm] non-reinforced membrane adhered to a ½ inch [13 mm] cover board mechanically attached through two loose-laid layers of 1½ inch [38 mm] polyisocyanurate insulation (3 inch [75 mm] total thickness) over the deck.
6. Mechanically attached PVC system: 45 mil [1.15 mm] membrane mechanically attached through a ½ inch [13 mm] cover board and two loose-laid layers of 1½ inch [38 mm] polyisocyanurate insulation (3 inch [75 mm] total thickness) over the deck.
7. SPF system: Silicone coating with granules over 3.2 inch [81 mm] minimum thickness of foam sprayed to ½ inch [13 mm] gypsum board mechanically attached to the deck.
8. Structural metal panel system: Attach galvanized steel Z members at 4 feet [1220 mm] on center to the deck with screws. Loose-lay two layers of 1¾ inch [44 mm] polyisocyanurate insulation (3½ inch [90 mm] total thickness) over the deck and between the Zs. Install self-adhering modified bitumen underlayment on top of the insulation. Place aluminum-zinc alloy steel (Galvalume) trapezoidal panel system over the underlayment.
   a. The underlayment provides moisture protection for the roof insulation prior to installation of the metal panels, and it provides secondary protection after panel installation. However, depending upon construction methods and seasonal conditions, the underlayment could be omitted.
   b. With a structural panel system, the steel deck could be eliminated, and less expensive batt insulation could be used. However, to simplify cost estimating, this approach was not taken.
   c. The differential between the BUR and metal panel systems could be greater than that shown in the table if the metal panel work is performed by sheet metal union labor. While the cost differential is minimal between union and non-union labor for other types of low-slope roofs, the difference between the union and non-union labor in some marketplaces is significant for sheet metal work.

Although there are difficulties and limitations with the LCC approach, economic decisions based on LCC are preferable to those that only consider initial system cost.

Warranty

Architects and building owners often give considerable weight to a manufacturer's warranty when considering a roof system and a specific manufacturer. As will be discussed in more detail in the next chapter, many limitations are associated with most warranties. The warranty itself should not be the basis for selecting a system or a manufacturer.

Consider the following real-life example of the pitfalls of warranty criteria overriding other selection criteria: An engineering firm was retained to evaluate wind damage to a roof, recommend a system for reroofing, and prepare specifications and drawings for the new roof. In selecting the replacement system, the engineer stated that consideration was given to the warranty, initial cost, maintenance cost and service life. He made no mention of the selected system's wind resistance. About a year after the new roof was installed, a major portion of the roof experienced another partial blow-off. That area was replaced. The following year, extensive wind damage occurred again and the entire building was reroofed again. Had the engineer devoted attention to wind, which caused the demise of the original roof, the outcome of the new roof would likely have been much different.

Implications of Sustainable Roof Design

If an emphasis on sustainable roof design is desired, sustainable design criteria can become major factors in the selection process, depending upon the degree to which sustainability is pursued. At the very least, the selected system should be thermally efficient, with consideration given to both R-value and reflectivity. And for those buildings that are intended to have a service life in excess of 20 years, a system with enhanced durability should be selected to reasonably maximize the life of the roof to the extent that the budget allows.

See the previous chapter for further information on sustainable design considerations.

MANUFACTURER SELECTION

After the system has been selected, often it is desirable to specify one or more manufacturers. It is recommended that the following be considered:

General Considerations

Product quality

The manufacturer's current products being considered should have a successful track record of at least five years. On some jobs, commodity (good) quality is desired; on others, top-of-the-line quality is expected. In either case, the manufacturer should have a good record of consistently producing the specified quality. Avoid manufacturers that periodically produce products below their normal quality level.

Technical support

The manufacturer should be capable of offering adequate support to the architect during design and to the architect and contractor during application. Support services include adequate product testing and adequate literature (such as complete test data, typical details, application instructions, and maintenance and repair instructions). Probably most important, the manufacturer should have an ample technical staff that fully understands the capabilities and limitations of its products and systems.

Distribution

Verify that the manufacturer distributes its products to the project location. Many manufacturers provide products throughout the United States; however, some are regional. If a manufacturer does not distribute into an area, it may not be worth the additional cost to ship in its products.

Problem resolution

It is important to select a manufacturer that is willing to assist in resolving a problem that arises during or after construction. This does not mean that the manufacturer should pay for a problem that is not theirs. But it does mean that the manufacturer should be willing to offer technical expertise in identifying the causes of the problem and to provide input into how the problem can be solved.

Some manufacturers are very good in assisting with problem resolution. Others are disruptive during the resolution process. They seek to blame others for the problem, and they offer little, if any, help in resolving it. If a problem develops, regardless the cause of the problem and who is at fault, it is very beneficial to be working with a manufacturer that provides good meaningful input.

System-specific Considerations

Metal roofing

If a metal roofing system is selected, consider specifying that the manufacturer be certified through the MBMA Metal Roofing Systems Quality Certification Program. This program examines and certifies the in-place capability of a manufacturer's organization and facilities to meet, and, on an ongoing basis, adhere to the program requirements regarding administrative policies, procedures, personnel qualifications, design procedures and practices, material procurement and usage, and manufacturing procedures, practices and quality control. The program is open to manufacturers that are not members of MBMA. Further information about the program is provided in MBMA's *Metal Roofing Systems Design Manual.*

SELECTION MATRIX

In "Roof Design by Weighted Evaluation," Luther C. Mock presents a system selection process called *weighted evaluation methodology*. The method can be used to help architects and their clients select a roof system. The evaluation is done in two steps. First, the most important project criteria are identified. Each criterion is compared with one another for relative importance and a score is assigned. Second, system alternatives are then selected and compared to each criterion and a score is assigned to each system. The scores of the alternatives are compared to see which system provides the highest score (*see a more detailed description and example in figure 115 and 115a, pages 100 and 101*).

It is critical, of course, that the scores assigned to a particular system in regard to its ability to accommodate the selected criteria be valid. For example, if the user of the method believes that a system is excellent (a score of 5) in accommodating a given criteria, but in reality the system is only fair (a score of 2), then the system that ranks the highest may indeed not be the best system. Hence, the methodology may be useful, but the inputs into scoring need to be rationally based. Unfortunately, in many instances, there are insufficient data to make a scoring decision without some subjectivity.

Architects interested in using the weighted evaluation methodology should obtain a complete copy of the paper. *Figure 116* is a blank copy of the worksheet.

FIGURE 115

Reprinted by permission from RCI.

ROOF SYSTEM DESIGN BY WEIGHTED EVALUATION

by Luther C. Mock

- **Step 1:** Identify and ask the question. For example, "What type of roof system should be installed?"
- **Step 2:** Brainstorm characteristics (criteria) that may be important factors relative to solutions (alternatives). By consensus, choose the most important criteria factors and list them on the criteria scoring matrix portion of the worksheet. See example worksheet.
- **Step 3:** Brainstorm solutions. By consensus, choose the best alternatives and list them on the analysis matrix portion of the worksheet.
- **Step 4:** Compare each criteria factor with each of the other criteria factors using the importance value scale of 1 to 4. Identify the more important of the two and assign an importance score. Take note in the example that "aesthetics" *(Item F)* was not considered a preference over any of the other six criteria factors; therefore, it did not receive any value.
- **Step 5:** Add the number of points that each criteria factor received and insert the total into the respective raw score cell of the matrix.
- **Step 6:** Identify the criteria factor with the largest raw score and assign a weight of importance factor of 10. Calculate each criteria factor's raw score as a direct relationship to a scale of 10. For example, if criteria factor "C" has the largest raw score of 16, it is assigned a weight of importance factor of 10. And, if criteria factor "A" has a raw score of 9, it is assigned a weight of importance of 6. Decimal values should be rounded off to the nearest whole number.
- **Step 7:** Compare each alternative from the analysis matrix to each criteria from the criteria scoring matrix using the analysis scoring scale of 1 to 5. Insert the value in the upper portion of the intersecting cell.
- **Step 8:** Multiply the analysis score by the corresponding criteria weight of importance factor to establish an adjusted sub-total score. Insert the value in the lower portion of intersecting cell.
- **Step 9:** Add each alternative's adjusted sub-total score to establish a final numerical score. The alternative with the largest numerical score is the preferred alternative (solution) to the problem.

FIGURE 115a

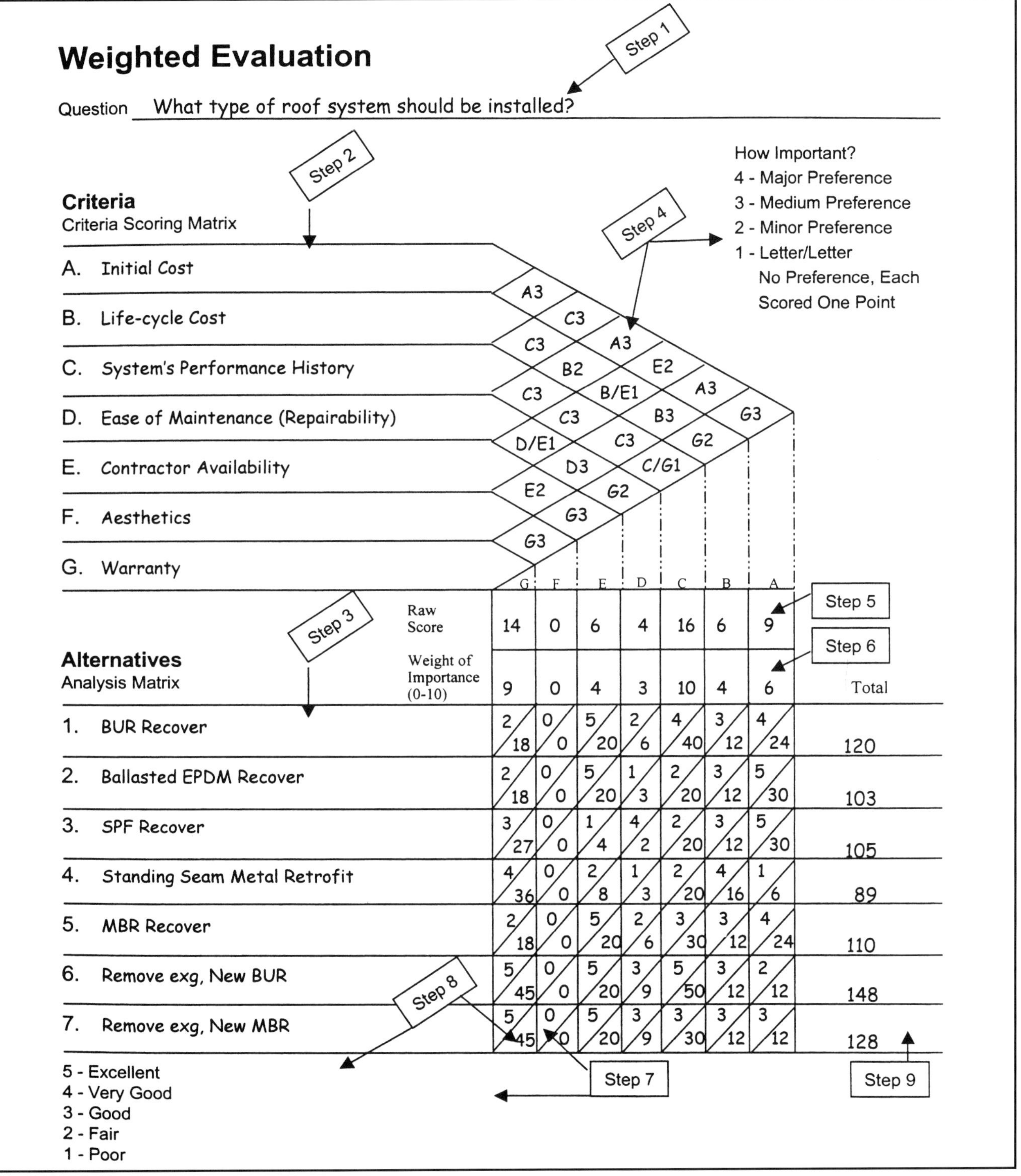

	G	F	E	D	C	B	A	Total
Raw Score	14	0	6	4	16	6	9	
Weight of Importance (0-10)	9	0	4	3	10	4	6	
1. BUR Recover	2 / 18	0 / 0	5 / 20	2 / 6	4 / 40	3 / 12	4 / 24	120
2. Ballasted EPDM Recover	2 / 18	0 / 0	5 / 20	1 / 3	2 / 20	3 / 12	5 / 30	103
3. SPF Recover	3 / 27	0 / 0	1 / 4	4 / 2	2 / 20	3 / 12	5 / 30	105
4. Standing Seam Metal Retrofit	4 / 36	0 / 0	2 / 8	1 / 3	2 / 20	4 / 16	1 / 6	89
5. MBR Recover	2 / 18	0 / 0	5 / 20	2 / 6	3 / 30	3 / 12	4 / 24	110
6. Remove exg, New BUR	5 / 45	0 / 0	5 / 20	3 / 9	5 / 50	3 / 12	2 / 12	148
7. Remove exg, New MBR	5 / 45	0 / 0	5 / 20	3 / 9	3 / 30	3 / 12	3 / 12	128

FIGURE 116

Blank Worksheet

Reprinted by permission from RCI.

Weighted Evaluation

Question ______________________________

How Important?
4 - Major Preference
3 - Medium Preference
2 - Minor Preference
1 - Letter/Letter
No Preference, Each
Scored One Point

Criteria
Criteria Scoring Matrix

A.

B.

C.

D.

E.

F.

G.

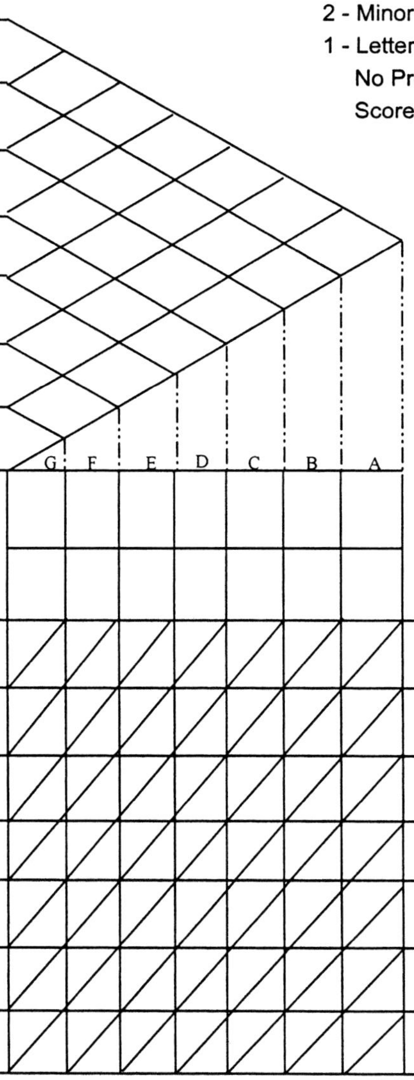

	G	F	E	D	C	B	A	
Raw Score								
Weight of Importance (0-10)								Total

Alternatives
Analysis Matrix

	G	F	E	D	C	B	A	Total
1.								
2.								
3.								
4.								
5.								
6.								
7.								

5 - Excellent
4 - Very Good
3 - Good
2 - Fair
1 - Poor

WARRANTY CONSIDERATIONS

Many building owners are concerned about the financial responsibility associated with premature roof failures. This concern is related to their perception that low-slope roofs frequently fail prematurely, and to the high cost of roof replacement and repairing the building's interior and contents due to leakage damage. Business interruption is also a concern with catastrophic failures, such as wind blow-off *(figure 117)* or hail damage. Owners and architects frequently attempt to deal with these concerns by requesting a warranty. A warranty by itself, however, is not necessarily an effective solution.

FIGURE 117

The metal roof blew off of this school, causing extensive interior water damage. The school was not useable for a considerable amount of time. Manufacturers' warranties typically do not cover such consequential damages.

When warranties are specified, the specification typically requires the warranty to be issued by the roof membrane manufacturer. It is normally specified that the warranty includes both the roof membrane materials (and perhaps other roof system components) and the roofing contractor's workmanship. Material-only warranties are also available from manufacturers. A manufacturer's warranty establishes a direct contractual relationship between the building owner and manufacturer. The length of coverage for a manufacturer's warranty is typically five to 25 years, with 10 years being the most common. A specification can also be written to require a warranty from the roofing contractor. The length of coverage for this type of warranty is normally two years.

A warranty is neither a maintenance contract nor is it an insurance policy. Furthermore, it does not ensure that leakage, damage caused by hail or wind, or another type of damage will not occur. Rather, a warranty defines specific legal rights and obligations of the owner and warrantor. It includes remedies and exclusions. Unfortunately, if the warrantor is out of business when the building owner experiences a problem, the warranty often becomes a useless piece of paper.

Over the past two decades, many warranties have become marketing tools rather than true reflections of demonstrated roof system performance. It is common to see 20-year warranties on roof membranes that have only been in the marketplace for a few years. When architects rely on warranties for performance, rather than pay attention to the factors that actually affect performance, the potential for premature failure dramatically increases.

BENEFITS OF WARRANTIES

A warranty may have some merit if it means that the manufacturer will take steps to minimize the potential for future problems (such as reviewing the architect's specification and details and providing meaningful inspection during application). Some manufacturers may not provide job-specific technical support on non-warranted jobs. However, as discussed in the next chapter, it is possible to obtain these services from manufacturers without specifying a warranty.

A warranty may also enhance the likelihood that a roof will be installed by a professional contractor. Most manufacturers that offer long-term warranties require that the application be performed by an authorized contractor. Some manufacturers limit the contractors that they authorize to install their products, and they provide those contractors with training and technical support. However, other manufacturers will authorize a contractor to install their products based on little more than the fact that the contractor has been awarded a job. In this case, rather than lose the job, the manufacturer adds another contractor to its list of approved applicators. Instead of relying on a warranty to obtain a qualified contractor, architects should specify contractor qualification requirements. This is discussed in more detail in the next chapter.

If a problem that is covered by the warranty occurs, and the warrantor is still in business, the presence of the warranty may lead to a quick resolution of the problem. Virtually every warranty issued by a manufacturer covers repair of leaks caused by defective materials and workmanship (if the warranty is not for materials-only) provided that the cause of the leakage is covered under the terms of the warranty. Without a warranty, the owner might have to pursue legal action to obtain relief. Pursuing legal action, however, may be too costly if the problem is small (for example, a few hundred or a few thousand dollars). Also, in some states, an owner cannot directly sue a manufacturer if the manufacturer did not issue a warranty. In these states, the owner would need to sue the general contractor, who would sue the roofing contractor, who would then sue the manufacturer. The presence of a warranty provides a direct avenue for the owner to pursue a claim with the manufacturer if the manufacturer does not respond to a problem covered under the warranty.

LIMITATIONS OF WARRANTIES

Warranties are normally prepared to limit the manufacturer's liability to a narrow scope of provisions rather than to provide protection for the building owner. Warranties typically preclude claims based on other theories of liability, including negligence and breach of contract. In addition, warranties typically exclude implied and express warranties established by the Uniform Commercial Code (UCC).

The UCC provides that goods shall be merchantable (in other words, they are fit for their particular purpose) in a contract for sale of goods. However, the length of coverage under the UCC is typically four years from the date of sale (some roof problems occur within four years but many do not). While the provisions under the UCC may be more favorable to the owner than the provisions in a manufacturer's warranty, the owner can obtain coverage for a much longer time with a manufacturer's warranty.

Most warranties contain several other provisions unfavorable to the owner. The most significant of these are as follows:

- the exclusion of consequential damage (including damage to building interiors and contents, and business interruption);
- the limitation of wind coverage to wind speeds that are typically well below the design wind speed prescribed in the building code;
- the exclusion of hail damage *(figure 118);*
- the limitation of leak repairs to patching the membrane rather than removing and replacing wet insulation;
- only the manufacturer can determine the applicability of the warranty; and
- the inclusion of several provisions, which are identified in the owner maintenance manuals co-produced by NRCA (ARMA/NRCA 1996a, NRCA/SPFD 1998a, SPRI/NRCA 1992), that could result in the building owner's inadvertent nullification of the warranty *(figure 119).*

FIGURE 118
Large hail ruptured this modified bitumen membrane, which was composed of a cap sheet and base sheet. Cap sheet granules were driven through the membrane and into the face of the insulation. Hail damage is typically excluded from manufacturers' warranties.

FIGURE 119
Vegetation roots can penetrate membranes, causing leaks. This type of leakage is typically excluded from manufacturers' warranties. With a minor amount of periodic maintenance, this type of problem can be eliminated.

RECOMMENDATIONS

Don't make a roof system or manufacturer selection on the basis of a warranty. Select a system for its suitability to the project and select a manufacturer that will provide quality materials, testing and technical support commensurate with project requirements as discussed in the previous chapter. Specify that the membrane manufacturer review the drawings and specifications, and inspect the application as discussed in the next chapter.

Do not automatically specify a warranty. Rather, discuss the benefits and limitations of a warranty with the building owner and recommend that he or she review this issue with their legal and insurance advisors to decide what is in his or her best interest. Refer to the *Consumer Advisory Bulletin*, published by NRCA in 1992 *(figure 120)*.

If specification of a warranty is being considered, review the warranty section of NRCA's *Low-Slope Roofing Materials Guide.* If the owner desires a warranty, recommend that the owner's legal advisor identify provisions of the manufacturer's standard warranty that are unacceptable, so that warranty provisions acceptable to the owner can be specified. The following are examples of changes to standard warranties that should be considered:

- delete warranty language that takes away owner rights
- delete warranty language that excludes other rights and remedies
- delete warranty language that sets a maximum dollar limit
- delete warranty language that prohibits transfer to a new owner
- delete or modify the unfavorable items listed above in the Limitations of Warranties section.

FIGURE 120

NRCA Bulletin on Warranties

Reprinted by permission from NRCA.

CONSUMER ADVISORY BULLETIN

ROOFING WARRANTIES

Current state-of-the-art low-slope roofing systems are the result of a century of research and innovation. The relatively recent introduction of numerous systems utilizing rubbers, plastics, modified asphalts and other synthetic materials caused manufacturers to focus attention upon warranties they offered and to employ long-term warranties as a marketing tool. The National Roofing Contractors Association (NRCA), in the interest of the roofing consumer, acknowledges the following concerns relative to manufacturers' warranties.

The length of a roofing warranty should not be the primary criterion in the selection of a roofing product or system because the warranty does not necessarily provide assurance of satisfactory roofing performance. The selection of a roofing system for a particular project application should be based upon the product's qualities and suitability for the prospective construction project. A long-term warranty may be of little value to a consumer if the roof does not perform satisfactorily and the owner is plagued by leaks. Conversely, if the roof system is well-designed, well-constructed and well-manufactured, the expense of purchasing a warranty may not be necessary.

Manufacturers who use long-term warranties as a marketing tool have encountered a highly competitive roofing market and have found themselves compelled to meet or exceed warranties of competitive manufacturers. It is suspected that is some cases the length of the warranty was established without appropriate technical research or documentation of in-place field performance.

Increased liability risk associated with long-term warranties has contributed to the recent demise of some manufacturers resulting in unanticipated and costly expenses for extensive roof repairs by roofing consumers. Unfortunately, there are a number of manufacturers who issued long-term warranties and who are no longer operating companies with the capability of honoring their warranty commitments, leaving consumers with an ineffective warranty and a serious roofing problem.

There is a common misconception by roofing consumers that long-term warranties are all-inclusive insurance policies designed to cover virtually any roofing problem, regardless of the cause or circumstance. Roof warranties typically do not warrant that the roof system will not leak or is suitable for the project where it is installed. Even the most comprehensive manufacturer warranties that cover material and workmanship generally provide only that the manufacturer will repair leaks that result from the specific causes enumerated in the warranty. A material-only warranty typically provides only that the manufacturer will provide replacement material.

Warranty documents often contain restrictive provisions which significantly limit the warrantor's liability and the consumer's remedies in the event that problems develop. The warranty document may also contain other restrictions and limitations such as a prohibition against assignment or transfer of the warranty, *exclusion* of damages resulting from a defective roof and monetary limitations.

Long-term warranties are largely reactive rather than pro-active solutions to roof problems. In general, they tend to undermine a prudent owner's initial concern for

FIGURE 120, (cont'd.)
NRCA Bulletin on Warranties
Reprinted by permission from NRCA.

proper roofing specifications and application, as well as their subsequent primary responsibility for periodic roof maintenance.

The roofing consumer is best served by:

- Manufacturers who focus their sales efforts primarily on the relevant and proven merits of those products and systems best designed to serve the specific needs of the roofing consumer.

- Manufacturers who base warranties for membranes or systems solely upon honest and realistic appraisal of their proven service life contingent upon the financial ability and good faith of the issuer to honor those warranties for the duration of the designated warranty term.

- Manufacturers who clearly and conspicuously state in writing all recommended, as well as required, owner maintenance responsibilities during the projected service life of the roof.

- Manufacturers who solicit from the roofing consumer a clear understanding of the consumer's primary responsibility to provide periodic routine maintenance during the service life of the roof membrane.

NRCA believes that the roofing consumer, with the assistance of a roofing professional, should focus his purchase decision primarily on a objective and comparative analysis of proven roofing system options that best serve his specific roofing requirements and not on warranty time frames.

NRCA further advises that the roofing consumer consult the membrane warranty section of the *Roofing Materials Guide* for a comparative analysis of the specific provision, remedies, limitations, and exclusions of the warranties of those roofing systems under consideration. All questions should be addressed to the respective roofing manufacturers for specific written clarification.

KEY ELEMENTS OF SPECIFICATIONS AND DRAWINGS

After a suitable roof system has been selected, appropriate specifications and drawings need to be prepared to help ensure that the architect's design concept and requirements are understood and executed by a professional roofing contractor. If a project with poorly prepared specifications or drawings is awarded to the low bidder, the likelihood that the contractor will perform quality work with high quality materials is small. The fate of prematurely failed roofs is typically set by poorly prepared documents *(figure 121)*. The importance of the architect's diligence in preparing specifications and drawings cannot be overemphasized.

FIGURE 121

This very wide gutter was not designed for wind uplift. All of the wind load was transferred to the edge nailer, but the nailer attachment did not account for this additional load. The nailer lifted, causing the roof membrane to peel back.

The following elements are critical in communicating project requirements to the contractor and ensuring delivery of a good roof to the building owner:

- specifications
- roof plan
- details
- peer review.

SPECIFICATIONS

To produce good specifications, the architect should first acquire a commercially available master guide specification (such as *MASTERSPEC*™). The following are important considerations when selecting such a guide.

- Technical competency: Does the master guide specification periodically go through extensive peer review from the roofing industry?
- Frequent updates: How frequently are sections updated? Fairly frequent substantive updates (about every three to four years) are important.
- Document coordination: Does the master guide specification include guidance on coordinating the specifications with the drawings?
- Background information: Does the master guide specification include information that helps the architect make decisions regarding the system being specified?

Manufacturers have guide specifications for their systems. These can be useful in tailoring a commercial master guide specification, but manufacturers' guide specifications should not serve as the architect's master.

After obtaining a commercial master guide specification, it is critical that the architect tailor it to the specific project. Information in the guide specification that is not applicable should be deleted. Applicable information should be adjusted if needed. Information should be added as necessary to suit the project. Architects should consider the following key elements when developing the roof specifications:

Part I - General

Building code requirements

Specify requirements for fire and wind uplift resistance. For wind uplift, specify the test method required to demonstrate required resistance.

Insurance requirements

If there are insurance requirements (such as FMG), specify what they are.

Submittals

Specify submittal of catalog data for all products, including installation instructions and, where applicable, maintenance and repair instructions. Specify submittal of samples only when necessary for the architect to evaluate the product. For example, it is normally not necessary to specify the submittal of an EPDM sample, because black sheets are typically specified and there is no factory-applied surfacing to evaluate. However, it may be appropriate to specify submittal of samples of a granule-surface SBS modified bitumen sheet in order to select the desired color and to qualitatively evaluate the granule coverage and embedment. A sample would usually not be necessary if a single proprietary SBS product is specified. But if the specification is open to several manufacturers, requiring a sample would be prudent.

Specify that the following be submitted: demonstration of specified contractor qualifications, manufacturer review letter, manufacturer inspection reports, certification reports demonstrating that materials comply with referenced standards, certificates of analysis (when specified), and documentation demonstrating enrollment in the MBMA certification program (when specified). These items are discussed in more detail below.

Contractor qualifications

Specify that the contractor has a minimum number of years of experience with the type of system specified (five years is usually a reasonable requirement). Also specify that the contractor is approved, authorized or licensed by the membrane materials manufacturer to install its product (although architects should realize that some manufactures require very little of a contractor to become one of their licensed applicators, as discussed in the previous chapter).

Manufacturer review

Require the membrane materials manufacturer to review the specification and drawings and advise in writing of its acceptance thereof, or to submit in writing its concerns and recommendations to resolve these concerns. Specify submittal of the review letter within a short time after contract award, so that if changes are needed there is sufficient time to develop them before the roofing begins.

Some manufacturers may not provide job specific technical support on non-warranted jobs. If a warranty is not specified, verify that the specified manufacturers will provide review and inspection (as discussed below). If the manufacturers will not provide these services, other manufacturers should be specified that will provide them, or specify a warranty (as discussed in previous chapter).

Manufacturer inspection
Require the membrane materials manufacturer to inspect the roof application on the first or second day of application, and to perform an inspection upon completion of the application. Require submittal of the inspection reports.

Standards compliance certificates
To help ensure that products furnished to the project comply with referenced standards, specify that certificates demonstrating compliance be submitted.

Certificates of analysis
Consider specifying certificates of analysis that provide quality control test results for specific manufacturing lots (lot numbers are printed on the product package label). Not all manufacturers currently offer this service.

Certification program
If a metal roofing system is selected, consider specifying that the manufacturer be certified through the MBMA Metal Roofing Systems Quality Certification Program *(see Manufacturer Selection in the "System Selection Criteria" chapter, page 97).*

Pre-roofing conference
Specify a two-stage pre-roofing conference. The first conference should be held a few weeks to several months prior to the start of roofing, depending upon job size and complexity. The second conference should be held just prior to application. The purpose of the conference is discussed in the next chapter.

Specify that the conference be attended by the general contractor, roofing contractor's project manager, superintendent (on large projects) and foreman, a representative of the membrane materials manufacturer, and the mechanical and electrical contractors if mechanical or electrical work is associated with the roofing work. If a third-party inspection firm will be on the job, the inspector should also attend.

Quality control documents
If a modified bitumen, single-ply or sprayed polyurethane foam system is selected, specify application compliance with the applicable quality-control document co-produced by NRCA (*see "Bibliography," page 159*).

Weather limitations
If there are weather-related limitations, they should be specified. If work will be performed during cold weather, special cold weather procedures should be specified *(figure 122)*.

Work hour limitations
If there are limitations to the hours or days that can be worked, they should be specified.

Storage areas
If there are limitations to on-site storage areas, they should be specified or shown on the drawings.

Load limits
Specify the maximum allowable wheel load limits for roof application equipment *(figures 123 and 124)*, and specify the maximum allowable load (per square foot [per square meter]) for materials stored on the roof. Some rooftop equipment and some materials, such as pallets of pavers, are very heavy. Specifying load limits will allow the contractor to determine the appropriate type of equipment and material handling and storage techniques needed for the project.

FIGURE 122
The roof on this building addition was applied during very cold weather. Warm air was blown into an air-supported enclosure that had been placed over the addition. The heater was placed outside of the enclosure, so that the moist combustion vapor was not exhausted into the work area.

FIGURE 123
A single-ply roll carrier is used to move large rolls of single-ply membrane across a roof. Depending upon the roll size, a single pair of tires can apply a load of up to around 1100 pounds [499 kg].

FIGURE 124
These 4-foot x 4-foot [1219-mm x 1219-mm] carts are loaded with mortar-faced XEPS boards. A fully loaded cart weighs about 1,000 pounds [453 kg]. Each wheel is carrying about 250 pounds [113 kg].

Maintenance contract

If the owner wishes to enter into a maintenance contract with the contractor for semi-annual inspection and routine maintenance, this should be specified.

Part II - Products

Materials

Avoid specifying unproven products. Specify that the manufacturer shall have produced the specified materials for a minimum of five years.

Materials are typically specified by listing a single manufacturer's specific products, or by listing products by several manufacturers, or by referencing a product standard (such as ASTM). Unfortunately, insufficient data currently exist to specify an entire system by performance criteria.

In the private sector, it is generally preferable to specify one or perhaps a few proprietary products for the membrane itself. If several manufacturers are listed, it is important for the architect to list those manufacturers and their products that are very similar to one another.

If products are specified by referencing a product standard (such as ASTM), it may be appropriate in some instances to just list the standard. But many ASTM standards have grades, types or both within the standard. In these cases, the architect needs to specify the type and grade. Also, additional information sometimes needs to be specified, such as product thickness. Products covered by some ASTM standards have a substantial range in physical properties. In some cases, it is appropriate to specify a physical property value that is different from that specified in the standard so that a higher quality is obtained. However, this approach should not be used as a method to exclude essentially equal products just because one product has a slightly different value.

Rather than specifying products by an ASTM product standard, products are sometimes specified by requiring certain physical properties based on a variety of ASTM test methods. While this can be a viable approach in those instances where there is no product standard, manufacturers sometimes advocate this approach when they want their products used at the exclusion of other manufacturers. In this scheme, the specification does not have the appearance of being proprietary, as only physical properties and ASTM test methods are specified. However, if only one manufacturer can meet the specified properties, it is a proprietary specification. This approach should only be used when no product standard is available or if the product standard is deemed inadequate. In these cases, the values and test methods should be rationally based, rather than developed for the purpose of surreptitiously eliminating competition.

Temporary roof

The architect should determine if a temporary roof is needed, as discussed in the "Reroofing Considerations" chapter. If it is deemed necessary, the architect should specify its material requirements in Part II and the execution requirements in Part III. It is inappropriate for the architect to specify something as vague as "provide a temporary roof if needed."

Part III - Execution

Work over or adjacent to a new or existing roof

When work will occur over or adjacent to a roof, special requirements should be specified as discussed in the "Design Considerations" and "Reroofing Considerations" chapters.

Protection of the completed work

If other construction work will occur over the new roof, or if other construction trades will be on the new roof, specify appropriate protection measures. In some instances, it is advisable to specify an additional roof inspection at the completion of the construction project to determine if damage occurred after completion of the roof.

Reroofing

Specification issues related to reroofing projects were previously addressed in the "Reroofing Considerations" chapter.

ROOF PLAN

The roof plan *(figure 125)* should be drawn to scale, with a note for the contractor to field verify the dimensions. The plan should be sufficiently large to adequately convey information. The plan should show all penetrations and all expansion, seismic and area divider joints. It should also show slope directions and approximate amount of slope. The plan should provide references to all penetrations, roof edges and roof-to-wall details. In addition, it should show the location of the different wind uplift areas (field, perimeter and corners).

On reroofing projects, a separate demolition plan and perhaps a phasing plan may be necessary.

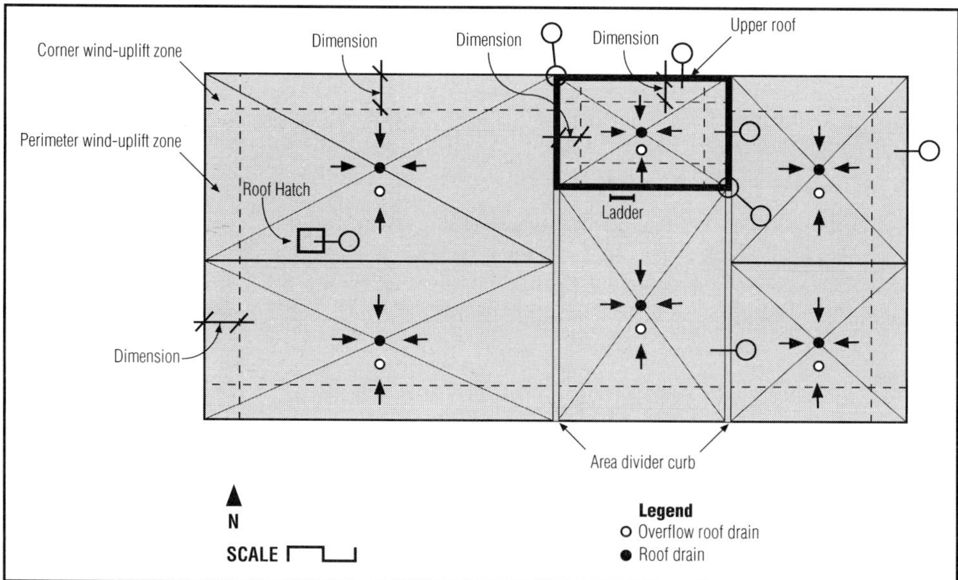

FIGURE 125
Roof Plan Example

DETAILS

Details should be drawn to scale and should be sufficiently large to adequately convey the information. Illustrating details in section typically suffices. However, sometimes an isometric drawing is needed in lieu of or in addition to a section. Typical and special conditions should be shown. Refer to the Flashings and Penetrations section of the "Design Considerations" chapter for more information (*page 51*).

At parapet and roof-to-wall details, the low point of the roof should be drawn and a dashed line placed at the high point if the roof elevation varies *(figure 126)*.

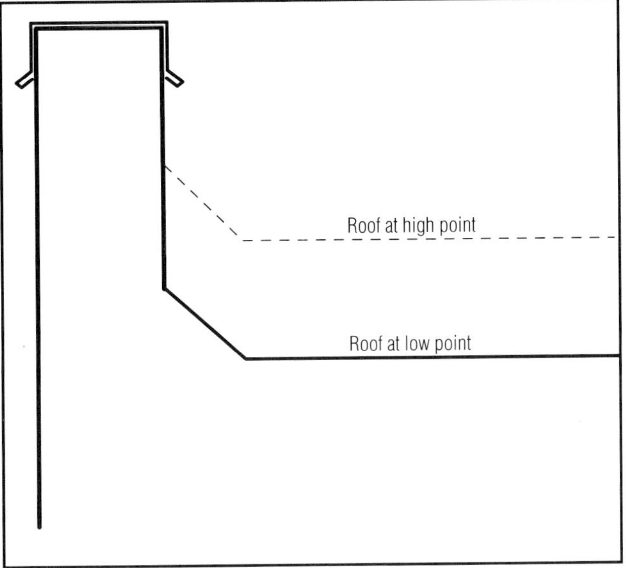

FIGURE 126
Roof-to-parapet Detail
Show both low and high points of roof.

PEER REVIEW

On some projects, it is prudent for the architect to have the specifications and drawings peer reviewed by someone knowledgeable of the specified system. Peer review should be considered for essential facilities, buildings with very valuable contents or operations, projects where the cost of the roofing work is very substantial, reroofing projects, complex or unusual projects, and those projects where the architect believes his or her expertise is lacking.

CONSTRUCTION CONTRACT ADMINISTRATION

During construction contract administration, the architect has several important tasks. Adequately executing these tasks is an important step in delivering a good roof to the building owner. These tasks need to be executed regardless of project size or location, although the amount of time devoted to the tasks will be dependent upon the roof size and other factors. If the building owner is unwilling to provide sufficient funds for the architect to adequately execute the construction contract administration tasks, the architect should seek to minimize his or her liability exposure.

Key elements of the following topics are discussed in this chapter:

- submittal review
- pre-roofing conference
- field observations
- project close-out.

SUBMITTAL REVIEW

Assuming that the specifications required a thorough submittal, it is incumbent upon the architect to be thorough, diligent and cautious during the submittal review process. (Note: In complex or unusual projects, or where the architect lacks sufficient expertise to adequately review the submittals, the architect should retain a professional roofing consultant who is knowledgeable about the particular roof system to review the submittals.) In particular, the architect should:

- Verify that all of the specified submittals are received and approved.
- If a submittal item is to be resubmitted, make sure that it is. Sometimes items to be resubmitted are forgotten about. Unfortunately these items often become problems.
- Be cautious in approving materials, systems and details that are not in accordance with the contract documents. Minor changes may affect code compliance or roof system performance.
- Prepare a memo identifying any changes to materials, systems or details and the reasons for the changes. If there are future problems, the memo could demonstrate that the architect acted responsibly in approving the changes.

PRE-ROOFING CONFERENCE

Assuming the specification requires a pre-roofing conference, it is important for the architect to work with the contractor to arrange for a conference at the appropriate time, and to remind the contractor that all parties listed in the specifications are required to attend. The purpose of the meeting is to review the drawings and specifications to ensure that there is understanding and agreement by all parties. If there are problems with the design or other aspects of the project, the intent is to identify and resolve them prior to commencement of the roofing work. The architect should prepare minutes for the conference and provide them to the contractor for distribution to the other parties.

If a two-stage conference was specified, the first conference should be held a few weeks to several months prior to the start of roofing, depending upon job size and complexity. The second conference should be held just prior to application. The advantage of two conferences is that the first can be held sufficiently in advance of application so that, if problems are discovered during the conference, there will be time to resolve them without delaying construction. Having the second conference just before application also provides an opportunity to verify that unresolved issues from the first conference have been dealt with, and it allows for a second review of critical issues just before commencement of work.

The agenda for both conferences is generally the same. Some items may be briefly covered in the first conference and then discussed more thoroughly in the second, or vice versa. Items that were discussed in detail in the first conference can often be quickly covered in the second meeting with brief mention and referral to the meeting minutes. The following should be undertaken at the conference:

- Review the salient features of the specifications, including schedule; demolition (if applicable); product delivery, storage and handling; roof loading; weather conditions; unique or critical items specified in Part III (Execution) of the specifications; and protection of the completed roofing from work of other trades.

  If it is a reroofing project, see the section entitled Special Aspects of Construction Contract Administration in the "Reroofing Considerations" chapter for additional discussion regarding unforeseen conditions (*page 83*).
- Review the drawings, particularly the unique or critical aspects.
- Review submittal problems (for example, items that have not been submitted, or items that have been submitted but rejected).
- Ensure that the contractor has a copy of the contract documents and approved submittals at the job site, and any changes thereto.
- Discuss the contractor's responsibilities regarding notification prior to roofing (for the purpose of alerting the field observer).
- Establish a line of communication between the field observer and contractor (and other parties that may be involved, such as the building owner's project manager or construction manager). If the field observer is not an employee of the architect, also establish a line of communication between the observer and the architect, and the extent of the authority that the observer has with respect to interpretation of the contract documents and the handling of unforeseen conditions.
- Immediately after the second conference, all attendees should review all of the roof deck areas to verify that they are ready for roofing. They should also review the parapets, curbs and penetrations. If a roof area is not ready, a later review of that area should be conducted prior to roofing. If corrective work is required, the corrective work should be reviewed prior to roofing.

FIELD OBSERVATIONS

For most projects, periodic observation by the architect is sufficient. But for others, full-time observation by the architect or a roof consultant is prudent. Building department inspectors should not be relied upon for quality assurance. The purpose of the observation is to help ensure that the work is being executed in accordance with the contract documents.

Amount of Observation

The amount of observation will depend upon:

Desired system reliability

If a highly reliable roof is desired, it should receive more field observation.

Characteristics of the roofing system

Some systems are more demanding, or less forgiving, than other systems with respect to workmanship and weather conditions at time of application. Even if the work is being performed by a knowledgeable and conscientious contractor, demanding systems present challenges. They should therefore receive greater observation in order to help avoid inadvertent mistakes.

For example, if a ribbon-applied foam adhesive is used to attach insulation or cover boards *(figure 127)*, increased observation is helpful. Because the adhesive is not applied over the entire face of the substrate, attention needs to be given to the ribbon spacing. If the spacing is excessive, the attachment could become overstressed during a windstorm. Also, where insulation boards are cut, the cutting debris should be removed prior to applying the adhesive. Because of the importance of a clean substrate with this type of adhesive attachment, an observer can help ensure that adequate attention is given to housekeeping *(figure 128)*.

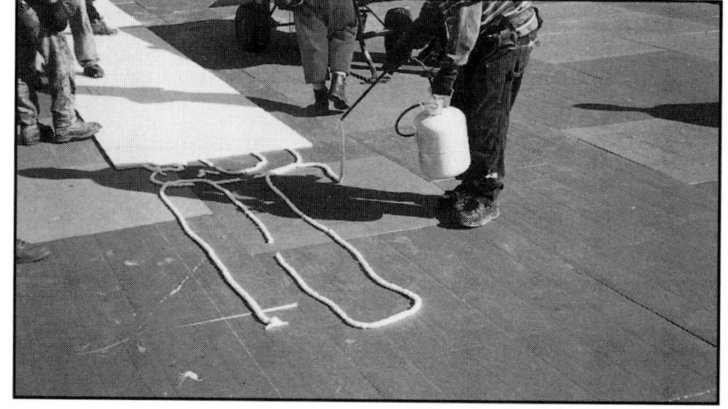

FIGURE 127
This cover board is being attached with a polyurethane foam adhesive. Increased field observation is prudent with this type of attachment method to ensure that the actual spacing between the adhesive ribbons is relatively close to that specified.

FIGURE 128
An insulation board was trimmed to fit in this corner area, but the trimming debris was not removed before application of the polyurethane foam adhesive. An observer can help ensure that the work area is clean before adhesive is applied.

Another example where increased observation is helpful is when a modified bitumen membrane is applied in cold adhesive when the ambient temperature is near the lower boundary recommended by the manufacturer. Increased observation could prevent the application from occurring if the temperature drops below the minimum recommended temperature.

Cost
As the cost of the roofing work increases, the amount of observation should also increase.

Complexity
Complex or unique roofs (for example, buildings with very high interior humidity or unusually shaped roofs) should receive greater observation.

Adjacency to existing roofs
Roofing work that occurs adjacent to or above an existing roof should typically receive greater observation to help ensure that the adjacent roof is not damaged by the other roofing work. See the Construction Above or Adjacent to a Roof section in the "Design Considerations" chapter (*page 59*).

Qualification of the roofing contractor
If the contractor is marginally qualified, greater observation should occur (full-time observation should be considered).

Reroofing
Reroofing projects typically should receive greater observation than new roof construction.

Role of the Observer

It is imperative that the observer thoroughly understands the system being installed. There is little value in having an observer who does not have the required knowledge. If the architect lacks sufficient expertise to provide the observation, a qualified professional roof consultant should be retained to perform this service. The observer should:

- Be provided with portions of the contract documents related to the roof.
- Be provided with a copy of the approved submittals.
- Be provided with copies of all changes related to the roof.
- Attend pre-roofing conferences.
- Verify that the materials on-site are those identified in the approved submittals, and that the materials have FM or UL labels when so specified.
- Follow the quality control guidelines co-produced by NRCA (ARMA/NRCA 1993, ARMA/NRCA 1996a, NRCA/SPFD 1998b, NRCA/SPRI 1997).
- If fastener pull-out tests were specified, verify that the results are acceptable. If the values are lower than anticipated, provide the contractor

with a revised fastening pattern that is commensurate with the test results *(figures 129 and 130).*

- If it appears that wheel loads or stored material loads exceed the specified load limits, the contractor should be advised immediately.
- Bring to the immediate attention of the contractor's job-site person (who was identified during the pre-roofing conference) any need for a change in the contractor's work practices or a need for corrective work *(figure 131).*
- As with submittal approval, be cautious in approving materials, systems and details that are not in accordance with the contract documents while performing field observations *(figure 132).* For example, the roofing crew may desire to use different fasteners to attach the membrane because the approved fasteners were not sent to the job site. Before accepting the fasteners, determine if wind uplift test ratings will be affected and if the membrane manufacturer will approve the alternative fasteners. Another example is a penetration detail that the foreman desires to flash differently than detailed. Is the change proposed in order to provide a better detail, or is it being proposed because

FIGURE 129

This mechanically attached EPDM membrane was experiencing problems before the job was finished. Many of the plastic auger fasteners had pulled out of the lightweight insulating concrete deck (*see figure 130*).

FIGURE 130

A test lab had performed pull-out tests. The pull-out results were well below what was required, but the lab mistakenly believed the results were adequate. This type of deck is not appropriate for a mechanically attached membrane system because it lacks resistance to the cyclical horizontal compression induced by the fasteners, as illustrated by the large hole worn by the fastener movements.

FIGURE 131

The dark area at the top of this SBS modified bitumen cap sheet is the selvage area. Without granules, this area will prematurely degrade. A strip of cap sheet needs to be added over the selvage area. An observer should be on the lookout for such mistakes.

FIGURE 132
This plumbing vent was not installed as detailed. The plumbing vent should have been centered on one of the standing seams. Instead it was centered between the seams. It becomes a dam in this location. To avoid mistakes such as this, this type of detail should be discussed during the pre-roofing conference, as it involves coordination between the mechanical and roofing subcontractors. Corrective action should be taken with this type of error.

it is simpler and cheaper to install? If the proposed detail is not as conservative as the original detail, it should probably not be approved.

- Write daily reports and give them expeditiously to the contractor (and to the architect if the observer is not an employee of the architectural firm that designed the roof). Report copies should be included in the file prepared for the building owner, as discussed in the following section.
- If it is a reroofing project, see the Special Aspects of Construction Contract Administration section in the "Reroofing Considerations" chapter for additional observation items (*page 83*).

PROJECT CLOSE-OUT

At the end of the project, the building owner should be provided a file with the following items:

- contract drawings and specifications related to the roof
- approved submittals
- minutes from the pre-roofing conferences
- field observation reports
- pertinent construction correspondence related to the roof
- warranty (if specified).

In addition to the above, it is prudent for the architect to furnish the building owner with one of the owner manuals co-produced by NRCA (ARMA/NRCA 1996a, NRCA/SPFD 1998a, SPRI/NRCA 1992). Owner manuals are not currently available for metal or liquid-applied roofs.

The owner should be advised that the roof should be inspected twice a year and that routine maintenance (such as debris removal) needs to be performed periodically. The owner should also be advised to keep the file current by adding to it the semi-annual inspection reports and documentation of any future leaks, repairs, additions or penetrations.

PROBLEMS AFTER JOB COMPLETION

If, after a new roof system has been installed and accepted, problems develop that threaten to diminish or abruptly end the roof's service life, the architect involved with the design of the roof system is often requested by the building owner to evaluate the problem. Also, architects are sometimes requested by building owners to investigate premature failure problems on roofs that they did not design. A request to investigate a premature failure may be received a few days or several years after acceptance of the roof.

This chapter discusses what an architect should do if requested to investigate and evaluate a premature failure. The following topics are included:

- types of premature failures
- investigation
- evaluation
- corrective action.

TYPES OF PREMATURE FAILURES

The most common premature failures are those that result in intermittent, modest or moderate water leakage into the building. These failures can be caused by poor design details, poor workmanship or abuse. If detected early, these types of problems can frequently be corrected relatively inexpensively. After proper corrective action, the roof system often can be expected to reach its intended service life.

An abrupt failure, on the other hand, allows a large quantity of water to enter the building in one or more isolated areas. An abrupt failure can be caused by a split in the membrane or a seam rupture in a single-ply membrane. A repair can be made, but if the problem is systemic (in other words, conditions are such that similar failures are likely to occur in several other areas of the roof), the roof has little chance of reaching its intended service life without periodic costly repairs and frustrating leaks. Other types of abrupt failures, such as those caused by hail or wind, often occur throughout the roof area. In these cases, roof replacement is frequently required in lieu of repair. Roof collapse is yet another type of abrupt failure. Although collapses are typically due to roof structure deficiencies, sometimes they can result from problems with the roof system (for example, roof drains clogged with debris and inadequate or inadequately located overflow drains or scuppers).

Premature degradation of the roof membrane is the least common cause of roof failure. In such cases, the rate of degradation sometimes can be slowed by the application of a roof coating after the degradation is discovered *(figure 133)*. Often, however, such membranes cannot be economically saved

FIGURE 133

This APP modified bitumen membrane did not have any protective surfacing. After a few years of exposure, significant surface cracking occurs. The roof service life can likely be prolonged by application of a coating.

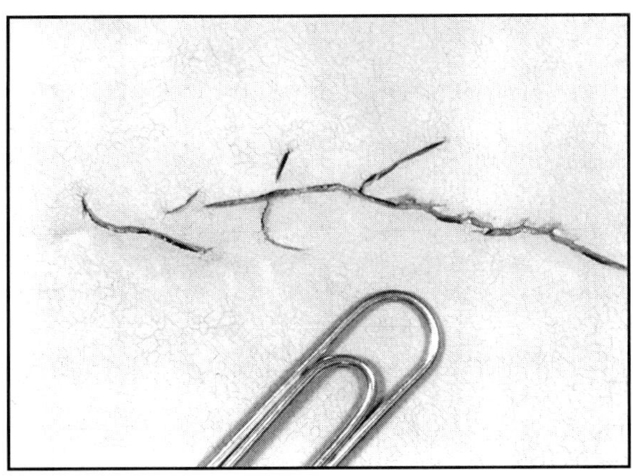

FIGURE 134
This 12-year-old reinforced PVC membrane has cracking that extends completely through the membrane. Similar cracks appeared in several other locations on the roof. A paper clip shows the scale. With this type of problem, it is unlikely that the roof can be economically saved.

(figure 134). Fortunately, with this kind of failure, there is usually ample time to evaluate the problem and plan appropriate action.

INVESTIGATION

If the roof is vulnerable to leakage or is actively leaking, the building owner frequently rushes to have the roof repaired or the building reroofed. If possible, however, an investigation should be undertaken and the appropriate documentation prepared prior to executing emergency repairs. A two-stage investigation is recommended. However, before conducting the investigation, various notifications should be made.

If the architect who designed the roof is contacted by the building owner to investigate the roof, the architect should notify his or her insurance company (except, perhaps, in the case of very minor problems). If the problem appears to be costly, it is also prudent for the architect to notify his or her attorney. The architect should advise the building owner to notify the roof membrane manufacturer, roofing contractor and general contractor so that those parties who were involved with the roof have an opportunity to perform their own investigation.

If an architect who did not design the roof is contacted by the building owner to investigate the roof, the architect should advise the building owner to notify the roof membrane manufacturer, roofing contractor, general contractor and the architect who designed the roof.

First Investigation

The first investigation is performed by the architect as soon as possible. If the failure is catastrophic, it is preferable to perform the investigation within minutes or hours after notification of the problem. The purpose of this investigation is to obtain baseline information on the magnitude of the roof damage and interior water damage. If the damage was caused by wind or if the roof collapsed, it should also be determined if anyone was injured and if adjacent property (such as other buildings, vehicles or landscaping) was damaged. Several high quality color photographs should be taken. The photography should be thorough, since these may be the only photos taken before the scene changes (due to emergency repairs).

In addition to the photos, video documentation is also recommended. To facilitate video analysis, pan the camera slowly and, when not panning, let each shot last several seconds. Describe conditions, such as direction of view, while shooting the video. *However, do not give your analysis of the problem.* The purpose of the video is to document conditions, not make assessments. Assessment is a very deliberative process that should be performed by a qualified investigator after collection of various pieces of background data.

Second Investigation

The purpose of the second-stage investigation is to thoroughly document the problem and damages, and to obtain data that can be used during the evaluation stage to determine the cause of the problem. Samples are often taken for field or laboratory analysis during this investigation stage *(figures 135 through 137).*

If the architect is not an experienced investigator and not knowledgeable about the type of roof system and type of existing problem, it is important to retain a qualified investigator to perform the second investigation. Misdiagnosis or poor documentation of the problem can occur if the second-stage investigation is not performed by a qualified investigator. This in turn can have very severe consequences, such as assignment of liability to the wrong party or failure to rectify the true cause of the problem.

For roofs designed by the architect, it is usually prudent to retain an independent investigator even if the architect is an experienced investigator to avoid any appearance of bias in the investigation.

FIGURE 135

The fasteners on this mechanically attached EPDM membrane were *tenting,* or protruding against the membrane like a tent pole. A sample was taken during the field investigation to determine why the tenting occurred (*see figure 136*).

FIGURE 136

The membrane shown in *figure 135* had been installed over ½-inch-thick [13 mm] wood fiberboard, which had been placed over an existing BUR. The wood fiberboard had become wet, which caused it to compress.

FIGURE 137

Leakage was experienced with this snap-together standing seam metal roof system. During the field investigation, a panel was removed and a layer of snow and ice was discovered between the panels and rigid insulation. Several openings through which snow could enter, including this improperly nested endlap, were found.

EVALUATION

After the field investigation, the next step is for the investigator to perform an evaluation of the problem. Sometimes the evaluation can be quickly performed and the course of corrective action soon implemented. Often, however, the evaluation takes considerable time. Furthermore, additional time for the evaluation will be necessary if some or all portions of the roof needs to be reroofed, and if the original design needs to be changed (as is often the case).

It is desirable to quickly reroof when the building is vulnerable to water infiltration. In these situations, however, the critical evaluation stage frequently does not receive the attention that it should. As a result, decisions can be made that result in a new roof that is no better (and sometimes worse) than the old one.

To avoid this problem, it is often a good idea to recommend that a temporary roof be installed after the investigation is complete. The temporary roof can provide months of protection while a thorough evaluation is made and an appropriate reroofing design developed.

CORRECTIVE ACTION

Permanent corrective action should only be taken after the problem has been adequately investigated and evaluated. Otherwise, the corrective action may itself fail prematurely. If the evaluation reveals that similar problems may develop outside of the original problem area, the corrective action should be executed in these other areas as well.

For further information on problems after job completion, refer to the article "The Roof System Blew Off—Now What?" by Thomas L. Smith. Although this reference pertains to wind blow-offs, the concepts are applicable to a wide variety of premature failures. This reference discusses emergency repairs, as do the following:

- *Manual of Roof Inspection, Maintenance, and Emergency Repair for Existing Single-Ply Roofing Systems* by SPRI and NRCA
- *Manual for Inspection and Maintenance of Built-Up and Modified Bitumen Roof Systems: A Guide for Building Owners* by ARMA and NRCA
- *Manual for Inspection and Maintenance of Spray Polyurethane Foam-Based Roof Systems: A Guide for Building Owners by* NRCA and SPFD.

ABBREVIATIONS

AAWE	American Association for Wind Engineering
AIA	The American Institute of Architects
APP	atactic polypropylene
ARMA	Asphalt Roofing Manufacturers Association
ASCE	American Society of Civil Engineers
ASHRAE	American Society of Heating, Refrigerating and Air-Conditioning Engineers
ASTM	American Society for Testing and Materials
Btu	British thermal unit
BUR	built-up roof(ing)
CADD	computer-aided design and drafting
CFC	chlorofluorocarbon
CGSB	Canadian General Standards Board
CIB	International Council for Research and Innovation in Building and Construction
CPA	copolymer alloy
CPE	chlorinated polyethylene
CRREL	U.S. Army Cold Regions Research and Engineering Laboratory
CSI	Construction Specifications Institute
CSM	chlorosulfonated polyethylene
CSPE	chlorosulfonated polyethylene
CR	neoprene
ECH	epichlorohydrin
EIP	ethylene interpolymer
EPA	U.S. Environmental Protection Agency
EPDM	ethylene propylene diene monomer
FEMA	Federal Emergency Management Agency
FM	Factory Mutual
FMG	Factory Mutual Global
FMR	Factory Mutual Research
HVAC	heating, ventilating and air-conditioning
IBC	*International Building Code*
KEE	ketone ethylene ester
kg	kilogram
kg/m^2	kilogram per square meter
kg/m^3	kilogram per cubic meter
kPa	kilopascal
LCC	life-cycle cost
m	meter
MB	modified bitumen
MBMA	Metal Building Manufacturers Association
MEPS	molded expanded polystyrene

ABBREVIATIONS

MH	moderate hail
MJ/m^2	megajoule per square meter
mm	millimeter
MRCA	Midwest Roofing Contractors Association
NBP	nitrile alloy
NCARB	National Council of Architectural Registration Boards
NDE	nondestructive evaluation
NRCA	National Roofing Contractors Association
ORNL	Oak Ridge National Laboratory
OSB	oriented strand board
PAO	polyalphaolefin
pcf	pounds per cubic foot
PDP	Professional Development Program
PIB	polyisobutylene
PMR	protected membrane roof
psf	pounds per square foot
psi	pounds per square inch
PVC	polyvinyl chloride
PVDF	polyvinylidene fluoride
RCI	Roof Consultants Institute
RIEI	Roofing Industry Educational Institute
RILEM	International Union of Testing and Research Laboratories for Materials and Structures
RRC	Registered Roof Consultant
SBCCI	Southern Building Code Congress International
SBS	styrene-butadiene-styrene
SEBS	styrene-ethylene-butylene-styrene
SH	severe hail
SIS	styrene-isoprene-styrene
SMACNA	Sheet Metal and Air Conditioning Contractors' National Association
SPF	sprayed polyurethane foam
SPFA	Spray Polyurethane Foam Alliance
SPFD	Sprayed Polyurethane Foam Division of The Society of the Plastics Industry, Inc.
SPRI	sheet membrane and component suppliers to the commercial roof industry (formerly the Single Ply Roofing Institute)
TPA	tripolymer alloy
TPO	thermoplastic polyolefin
UCC	Uniform Commercial Code
UL	Underwriters Laboratories Inc.
UV	ultraviolet radiation
XEPS	extruded expanded polystyrene

TERMS

The following terms are reprinted by permission of NRCA. They were taken from the glossary of the 5th edition of The NRCA Roofing and Waterproofing Manual. *The NRCA glossary also includes terms related to steep-slope roofs, waterproofing and general construction; these terms, however, are not listed below.*

Acrylic Coating
A liquid coating system based on an acrylic resin. Generally, a latex-based coating system that cures by air drying.

Aggregate
(1) Crushed stone, crushed slag or water-worn gravel used for surfacing a built-up roof system; (2) any granular material.

Aged R-value
Thermal resistance value established by utilizing artificial conditioning procedures for a prescribed time period.

Alligatoring
The cracking of the surfacing bitumen on a bituminous roof or coating on a SPF roof, producing a pattern of cracks similar to an alligator's hide; the cracks may not extend completely through the surfacing bitumen or coating.

Aluminized Steel
Sheet steel with a thin aluminum coating bonded to the surface to enhance weathering characteristics.

Architectural Panel
A metal roof panel, typically a double standing seam or batten seam; usually requires solid decking underneath and relies on slope to shed water.

Area Divider
A raised, flashed assembly, typically a single- or double-wood member attached to a wood base plate, that is anchored to the roof deck. It is used to accommodate thermal stresses in a roof system where an expansion joint is not required, or to separate large roof areas or separate roof systems comprising different/incompatible materials, and may be used to facilitate installation of tapered insulation.

Asphalt
A dark brown or black substance found in a natural state or, more commonly, left as a residue after evaporating or otherwise processing crude oil or petroleum. Asphalt may be further refined to conform to various roofing grade specifications:

> Dead-level asphalt: a roofing asphalt conforming to the requirements of ASTM Specification D 312, Type I.

Flat asphalt: a roofing asphalt conforming to the requirements of ASTM Specification D 312, Type II.

Steep asphalt: a roofing asphalt conforming to the requirements of ASTM Specification D 312, Type III.

Special steep asphalt: a roofing asphalt conforming to the requirements of ASTM Specification D 312, Type IV.

Asphalt Emulsion
A mixture of asphalt particles and emulsifying agent, such as bentonite clay and water.

Asphalt Felt
An asphalt-saturated and/or asphalt-coated felt.

Asphalt Primer
(*see Primer*)

Asphalt Roof Cement
A trowelable mixture of solvent-based bitumen, mineral stabilizers, other fibers and/or fillers. Classified by ASTM Standard D 2822-1 Asphalt Roof Cement, and D 4586-2 Asphalt Roof Cement, Asbestos-Free, Types I and II.

Type I is sometimes referred to as "plastic cement," and is made from asphalt characterized as self-sealing, adhesive and ductile, and conforming to ASTM Specification D 312, Type I; Specification D 449, Types I or II; or Specification D 946. *(see Plastic Cement and Flashing Cement)*

Type II is generally referred to as "vertical-grade flashing cement," and is made from asphalt characterized by a high softening point and relatively low ductility, and conforming to the requirement of ASTM Specification D 312, Types II or III; or Specification D 449, Type III. (*see Plastic Cement and Flashing Cement)*

Atactic Polypropylene
A group of high molecular weight polymers formed by the polymerization of propylene.

Ballast
A material, such as aggregate or precast concrete pavers, which employs its mass and the force of gravity to hold (or assist in holding) single-ply roof membranes in place.

Base Flashing (Membrane Base Flashing)
Plies or strips of roof membrane material used to close-off and/or seal a roof at the horizontal-to-vertical intersections, such as at a roof-to-wall juncture. Membrane base flashing covers the edge of the field membrane. (*see Flashing*)

Base Ply
The bottom or first ply in a built-up roof membrane when additional plies are to be subsequently installed.

Base Sheet
An impregnated, saturated or coated felt placed as the first ply in some low-slope roof systems.

Batten
(1) Cap or cover; (2) in a metal roof, a metal closure set over, or covering the joint between, adjacent metal panels; (3) in a wood roof, a strip of wood usually set in or over the structural deck, used to elevate and/or attach a primary roof covering such as tile; (4) in a single-ply membrane roof system, a narrow plastic, wood or metal bar that is used to fasten or hold the roof membrane and/or base flashing in place.

Batten Seam
A metal panel profile attached to and formed around a beveled wood or metal batten.

Bird Bath
Random, inconsequential amounts of residual water on a roof membrane.

Bitumen
(1) A class of amorphous, black or dark colored (solid, semi-solid or viscous) cementitious substances, natural or manufactured, composed principally of high molecular weight hydrocarbons, soluble in carbon disulfide, and found in asphalts, tars, pitches and asphaltenes; (2) a generic term used to denote any material composed principally of bitumen, typically asphalt or coal tar.

Bitumen-stop
(*see Envelope or Bleed-sheet*)

Bituminous Emulsion
A suspension of minute particles of bituminous material in water.

Bleed-sheet
A sheet material used to prevent the migration of bitumen.

Blister
An enclosed pocket of air, which may be mixed with water or solvent vapor, trapped between impermeable layers of felt or membrane, or between the membrane and substrate.

Boot
(1) A covering made of flexible material, which may be preformed to a particular shape, used to exclude dust, dirt, moisture, etc., from around a penetration; (2) a flexible material used to form a closure, sometimes installed at inside and outside corners.

Bridging
(1) When membrane or base flashing is unsupported at a juncture; (2) bridging in steep-slope roofing occurs when reroofing over standard-sized asphalt shingles with metric-sized asphalt shingles.

Broadcast
Uniformly cast or distribute granular or aggregate surfacing material.

Brooming
To improve the embedding of a ply or membrane by using a broom or squeegee to smooth it out and ensure contact with the adhesive under the ply or membrane.

Buckle
An upward, elongated displacement of a roof membrane frequently occurring over insulation or deck joints. A buckle may be an indication of movement within the roof assembly.

Built-up Roof (BUR)
A continuous, semi-flexible roof membrane, consisting of multiple plies of saturated felts, coated felts, fabrics or mats assembled in place with alternate layers of bitumen, and surfaced with mineral aggregate, bituminous materials, a liquid-applied coating or a granule-surfaced cap sheet.

Butyl
Rubber-like material produced by polymerizing isobutylene.

Butyl Coating
An elastomeric coating system derived from polymerized isobutylene. Butyl coatings are characterized by low water vapor permeability.

Butyl Rubber
A synthetic elastomer based on isobutylene and a minor amount of isoprene. It can be vulcanized and features low permeability to gases and water vapor.

Butyl Tape
A sealant tape sometimes used between metal roof panel seams and/or end laps; also used to seal other types of sheet metal joints and in various sealant applications.

Cant
In SPF-based roofing, a beveling of foam at horizontal/vertical joints to increase strength and promote water run off.

Cant Strip
A beveled strip used under flashings to modify the angle at the point where the roofing or waterproofing membrane meets any vertical element.

Cap Flashing
(1) Usually composed of metal, used to cover or shield the upper edges of the membrane base flashing wall flashing; (2) a flashing used to cover the top of various buildings components, such as parapets or columns. (*see Flashing*)

Cap Sheet
A sheet, often granule-surfaced, used as the top ply of some built-up or modified bitumen roof membranes and/or flashings.

Capacitance Meter
A device used to locate moisture or wet materials within a roof system by measuring the ratio of the change to the potential difference between two conducting elements separated by a non-conductor.

Chlorinated Polyethylene (CPE)
A thermoplastic material, used for single-ply roof membranes, composed of high molecular weight polyethylene that has been chlorinated with a process that yields a flexible rubber-like material.

Chlorosulfonated Polyethylene (CSPE or CSM)
Probably best known by the DuPont trade name Hypalon™, a synthetic, rubber-like thermoset material, based on high molecular weight polyethylene with sulphonyl chloride, usually formulated to produce a self-vulcanizing membrane. Classified by ASTM Standard D 5019.

Cleat
A continuous metal strip, or angled piece, used to secure metal components.

Coal Tar
A dark brown to black colored, semi-solid hydrocarbon produced by the distillation of coal. Coal tar pitch is further refined to conform to the following roofing grade specifications:

Coal tar pitch: a coal tar used as the waterproofing agent in dead-level or low-slope built-up roof membranes and membrane waterproofing systems, conforming to ASTM Specification D 450, Type I.

Coal tar waterproofing pitch: a coal tar used as the dampproofing or waterproofing agent in below-grade structures, conforming to ASTM Specification D 450, Type II.

Coal tar bitumen: a proprietary trade name for Type III coal tar used as the dampproofing or waterproofing agent in dead-level or low-slope built-up roof membranes and membrane waterproofing systems, conforming to ASTM D 450, Type III.

Coal Tar Felt
A felt that has been saturated or impregnated with refined coal tar.

Coal Tar Roof Cement
A trowelable mixture of processed coal tar base, solvents, mineral fillers and/or fibers. Classified by ASTM Standard D 4022, "Coal Tar Roof Cement, Asbestos Container."

Coarse Orange Peel Surface Texture
A surface showing a texture where nodules and valleys are approximately the same size and shape. This surface is acceptable for receiving a protective coating because of the roundness of the nodules and valleys.

Coated Base Sheet
A coated felt intended to be used as a base ply in a built-up or modified bitumen roof membrane.

Coating
A layer of liquid material applied to a surface for protection or appearance.

Coil Coating
The application of a finish to a coil of metal using a continuous mechanical coating process.

Cold Roof Assembly
A roof assembly configured with the insulation below the deck, not typically in contact with the deck, allowing for a ventilation space. The temperature of the roof assembly remains close to the outside air temperature.

Compounded Thermoplastics
A category of roofing membranes made by blending thermoplastic resins with plasticizers, various modifiers, stabilizers, flame retardants, UV absorbers, fungicides and other proprietary substances alloyed with proprietary organic polymers.

Conductance, Thermal
The thermal transmission in unit time through unit area of a particular body or assembly having defined surfaces, when unit average temperature difference is established between the surfaces. $C=Btu/h\ ft^2F$ [$C=W/m^2K$].

Contact Cements
Adhesives used to adhere or bond various roofing components. These adhesives adhere mated components immediately on contact of surfaces to which the adhesive has been applied.

Copolymer
The product of polymerization of two or more substances (as two different isomers) together.

Copolymerization
A chemical reaction that results in the bonding of two or more dissimilar monomers to produce large, long-chain molecules that are copolymers.

Core Cut or Core Sample
(1) A sample from a low-slope roof system taken for the purpose of obtaining primarily qualitative information about its construction. Typically, core cut analysis can verify or reveal the type of membrane surfacing; the type of membrane; the approximate number of plies; the type, thickness and condition of the insulation (if any); and the type of deck used as a substrate for the roof system. (2) For SPF-based roof systems, core cuts are used to obtain both quantitative and qualitative information, such as the thickness of the foam, the thickness and adhesion of the coating, thickness of individual passes and adhesion between passes, and the adhesion of the foam to its substrate.

Counter Batten
Vertical wood strips installed on sloped roofs over which horizontal battens are secured. The primary roof covering is attached or secured to these horizontal battens.

Counterflashing
Formed metal or elastomeric sheeting secured on or into a wall, curb, pipe, rooftop unit or other surface, to cover and protect the upper edge of a base flashing and its associated fasteners.

Course
(1) The term used for a row of roofing material that forms the roofing, waterproofing or flashing system; (2) one layer of a series of materials applied to a surface (e.g., a five-course wall flashing is composed of three applications of roof cement with one ply of felt or fabric sandwiched between two layers of roof cement).

Cover Board
An insulation board used over closed cell plastic foam insulation (e.g., polyisocyanurate) to prevent blistering when used in conjunction with hot bituminous membranes. Suitable cover board insulation are glass-faced siliconized gypsum board, glass-fiber board, perlite board, wood-fiber board or mineral-fiber board. Cover boards are also recommended between polyisocyanurate insulation and single-ply membranes to protect the polyisocyanurate.

Cover Plate
A metal strip sometimes installed over or under the joint between formed metal pieces.

Cream Time
Time in seconds (at a given temperature) when the A and B components of polyurethane foam will begin to expand after being mixed. Recognizable as a change in color of the materials.

Cricket
A relatively small area of a roof constructed to divert water from a horizontal intersection of the roof with a chimney, wall, expansion joint or other projection. (*see Saddle*)

Cross-linking
The formation of chemical bonds between polymeric chains. Cross-linking of rubber is referred to as vulcanization or "curing."

Cure
A process whereby a material is caused to form permanent molecular linkages by exposure to chemicals, heat, pressure and/or weathering.

Cutback
Solvent-thinned bitumen used in cold-process roofing adhesives, roof cements and roof coatings.

Cutoff
A permanent detail designed to prevent lateral water movement in an insulation system and used to isolate sections of a roofing system. (Note: A cutoff is different from a tie-in, which may be a temporary or permanent seal.) (*see Tie-in*)

Dead-level Asphalt
(*see Asphalt*)

Dew-point Temperature
The temperature at which air becomes saturated with water vapor. The temperature at which air has a relative humidity of 100 percent.

Double Lock Standing Seam
In a metal roof panel or metal cap, a standing seam that uses a double overlapping interlock between two metal panels. (*see Standing Seam*)

Drip Edge
A metal flashing or other overhanging component with an outward projecting lower edge, intended to control the direction of dripping water and help protect underlying building components.

Dry Film Thickness
The thickness, expressed in mils, of an applied and cured coating or mastic. (*For comparison, see Wet Film Thickness.*)

Dual Level Drain
In waterproofing, an outlet or other device with provisions for drainage at both the wearing surface and waterproofing membrane levels used to collect and direct the flow of runoff water from a horizontal slab.

Edge Stripping
Membrane flashing strips cut to specific widths used to seal/flash perimeter edge metal and the roof membrane application of felt strips cut to narrower widths than the normal felt-roll width to cover a joint between metal perimeter flashing and built-up roofing.

Edge Venting
The practice of providing regularly spaced or continuously protected (e.g., louvered) openings along a roof edge or perimeter, used as part of a ventilation system to dissipate heat and moisture vapor.

Elastomer
A macromolecular material that returns rapidly to its approximate initial dimensions and shape after substantial deformation by a weak stress and subsequent release of that stress.

Elastomeric Coating
A coating that is capable of being stretched at least twice its original length (100 percent elongation) and recovering to its original dimensions.

Embedment
(1) The process of pressing/positioning a felt, aggregate, fabric, mat or panel into hot bitumen or adhesive to ensure intimate contact at all points; (2) the process of pressing/positioning granules into coating in the manufacture of factory-prepared roofing, such as shingles.

Emulsion
A mixture of bitumen and water, with uniform dispersion of the bitumen or water globules, usually stabilized by an emulsifying agent or system.

Envelope (Bitumen-stop)
A continuous membrane edge seal formed at the perimeter and at penetrations by folding the base sheet or ply over the plies above and securing it to the top of the membrane. The envelope prevents bitumen seepage from the edge of the membrane.

Epichlorohydrin (ECH)
A synthetic rubber including two epichlorohydrin based elastomers. It is similar to and compatible with EPDM.

Epoxy
A class of synthetic, thermosetting resins that produce tough, hard, chemical-resistant coatings and adhesives.

Equilibrium Moisture Content (EMC)
(1) The moisture content of a material stabilized at a given temperature and relative humidity, expressed as percent moisture by weight.

Equiviscous Temperature (EVT)
The temperature at which a bitumen attains the proper viscosity for built-up membrane application.

Equiviscous Temperature (EVT) Application Range
The recommended bitumen application temperature range. The range is approximately 25 F [14 C] above or below the EVT, thus giving a range of approximately 50 F [28 C]. The EVT range temperature is measured in the mop cart or mechanical spreader just prior to application of the bitumen to the substrate. The recommended equiviscous temperature (EVT) for roofing asphalt (ASTM D 312, Type I, II, III or IV) is as follows:

> Mop application: the temperature at which the asphalt's apparent viscosity is 125 centipoise [0.125 Pa·s].
>
> Mechanical spreader application: the temperature at which the asphalt's apparent viscosity is 75 centipoise [0.075 Pa·s].
>
> Note: In order to avoid the use of two kettles if there are simultaneous mop and mechanical spreader applications, the EVT for mechanical spreader application can be used for both application techniques.

Ethylene Interpolymers (EIP)
A group of thermoplastic compounds generally based on PVC polymers from which certain single-ply roofing membranes can be formulated.

Ethylene Propylene Diene Terpolymer (EPDM)
Designated nomenclature of ASTM for a terpolymer of ethylene, propylene and diene. EPDM material is a thermosetting synthetic elastomer.

Expansion Cleat
A cleat designed to accommodate thermal movement of metal roof panels.

Exposure
(1) The traverse dimension of a roofing element or component not overlapped by an adjacent element or component in a roof covering. For example, the exposure of any ply in a built-up roof membrane may be computed by dividing the felt width, minus 2 inches [51 mm], by the number of shingled plies; thus, the exposure of 36-inch [914-mm] wide felt in a shingled, four-ply membrane should be approximately 8½ inches [216 mm]; (2) the dimension of sidewall or roofing covering that is not covered or overlapped by the up slope course of component. The typical exposure for a standard-sized, three-tab shingle is 5 inches [127 mm], depending on manufacturer specifications.

Fabric
A woven cloth or material of organic or inorganic filaments, threads, or yarns used for reinforcement in certain membranes and flashings.

Factory Seam
A splice/seam made by the manufacturer during the assembly of sections of materials into larger sheets/panels.

Field Seam
A splice or seam made in the field (not factory) where overlapping sheets are joined together using an adhesive, splicing tape, or heat- or solvent-welding.

Film
Sheeting having a nominal thickness not greater than 10 mils [0.25 mm].

Film Thickness
The thickness of a membrane or coating. Wet film thickness is the thickness of a coating as applied; dry film thickness is the thickness after curing. Film thickness is usually expressed in mils (thousandths of an inch).

Fine Mineral-surfacing
Water-insoluble, inorganic material, more than 50 percent of which passes through a No. 35 sieve. Used on the surface of various roofing materials and membranes to prevent sticking.

Flame Spread
The propagation of a flame away from its source of ignition.

Flash Point
The lowest temperature at which vapors above a volatile combustible substance ignite in air when exposed to a flame.

Flashing
Components used to weatherproof or seal roof system edges at perimeters, penetrations, walls, expansion joints, valley, drains and other places where the roof covering is interrupted or terminated. For example, membrane base flashing covers the edge of the field membrane, and cap flashings or counterflashings shield the upper edges of the base flashing.

Flashing Cement
A trowelable mixture of solvent-based bitumen and mineral stabilizers that may include asbestos or other inorganic or organic fibers. Generally, flashing cement is characterized as vertical-grade, which indicates it is intended for use on vertical surfaces. (*see Asphalt Roof Cement and Plastic Cement*)

Foam Stop
The roof edge treatment upon which SPF is terminated.

Froth Pack
A term used to describe small, disposable aerosol cans that contain SPF components. Two component froth packs are available to do small repairs for sprayed polyurethane foam-based roofs.

Glass Mat
A thin mat of glass fibers with or without a binder.

Granule (also referred to as Mineral or Ceramic Granule)
Opaque, natural or synthetically colored aggregate commonly used to surface cap sheets, shingles and other granule-surfaced roof coverings.

Heat Seaming
The process of joining thermoplastic films, membranes or sheets by heating and then applying pressure to bring both materials in contact with each other (*see Heat Welding*).

Heat Welding
Method of melting and fusing together the overlapping edges of separate sheets or sections of polymer modified bitumen, thermoplastics or some uncured thermoset roofing membranes by the application of heat (in the form of hot air or open flame) and pressure. (*see Heat Seaming*)

Hem
The edge created by folding metal back on itself.

Holiday
An area where a liquid-applied material is missing or absent.

Hot or Hot Stuff
A roofing worker's term for hot bitumen.

Hybrid Roof Covering
Combination of two or more separate and distinct roof membranes; e.g., three-ply smooth BUR and a modified bitumen cap.

Hypalon™
A registered trademark of E.I. du Pont de Nemours & Co., for "chlorosulfonated polyethylene," or CSPE. (*see Chlorosulfonated Polyethylene*)

Ice Dam
A mass of ice formed at the transition from a warm to a cold roof surface, frequently formed by refreezing meltwater at the overhang of a steep roof, causing ice and water to back up under roofing materials.

Infrared Thermography
The process of displaying variations of apparent temperatures (variation of temperature or emissivity or both) over the surface of an object by measuring variations in infrared radiance.

In-service R-value
Thermal resistance value established under installed conditions and measured over the expected service life of the material.

Inverted Roof Membrane Assembly (IRMA™)
A patented, proprietary variation of the "protected membrane roof assembly" in which Styrofoam® brand insulation and ballast are placed over the roof membrane. IRMA™ and Styrofoam® are registered trademarks of the Dow Chemical Company. [*NCARB editor's note: Although protected membrane roofs are sometimes referred to as "inverted roofs," the trademark "IRMA" stands for "Insulated Roof Membrane Assembly."*]

Isocyanate
A highly reactive organic chemical containing one or more isocyanate (-N=C=0) groups. A basic component in SPF-based systems and some polyurethane coating systems.

Isolation Sheet
(*see Slip Sheet*)

Lap Seam
Occurs where overlapping materials are seamed, sealed or otherwise bonded.

Latex
A stable dispersion of polymeric substance in an essentially aqueous medium.

Lift
The sprayed polyurethane foam that results from a pass. It usually is associated with a certain pass thickness and has a bottom layer, center mass and top skin in its makeup.

Liquid-applied
Application of bituminous cements, adhesives or coatings installed at ambient or slightly elevated temperatures.

Loose-laid Membrane
A ballasted roofing membrane that is attached to the substrate only at the edges and penetrations through the roof.

Low-slope Roofs
A category of roofs that generally includes weatherproof membrane types of roof systems installed on slopes at or less than 3:12 (14 degrees).

Low Temperature Flexibility
The ability of a membrane or other material to resist cracking when flexed after it has been cooled to a low temperature.

Mat
A thin layer of woven, non-woven or knitted fiber that serves as reinforcement to a material or membrane.

Mechanically Fastened Membranes
Generally used to describe membranes that have been attached at defined intervals to the substrate.

Membrane
A flexible or semi-flexible roof covering or waterproofing whose primary function is to exclude water.

Metal Roof Panel
An interlocking metal sheet having a minimum installed weather exposure of 3 square feet [279,000 mm^2 or 0.28 m^2] per sheet.

Mil
A unit of measure, one mil is equal to 0.001 inches, or 25.4 micrometers (mm), often used to indicate the thickness of a roofing membrane.

Mineral Granules
(*see Granule*)

Mineral Stabilizer
A fine, water-insoluble inorganic material, used in a mixture with solid or semi-solid bituminous materials.

Mineral-surfaced Sheet
A roofing sheet that is coated on one or both sides with asphalt and surfaced with mineral granules.

Modified Bitumen
(1) A bitumen modified by including one or more polymers (e.g., atactic polypropylene, styrene-butadiene-styrene, etc.); (2) composite sheets consisting of a polymer modified bitumen often reinforced with various types of mats or films and sometimes surfaced with films, foils or mineral granules.

Moisture Contour Map
A map used to graphically define the location of moisture within a roof assembly after a moisture scan has been performed.

Moisture Relief Vent
A venting device installed through the roofing membrane to relieve moisture vapor pressure from within the roofing system.

Moisture Scan
The use of a mechanical device (capacitance, infrared or nuclear) to detect the presence of moisture within a roof assembly. (*see Nondestructive Testing*)

Mole Run
A meandering ridge in a roof membrane not associated with insulation or deck joints.

Monomer
A low-molecular-weight substance consisting of molecules capable of reacting with like or unlike molecules to form a polymer.

Mop-and-flop
An application procedure in which roofing elements (insulation boards, felt plies, cap sheets, etc.) are initially placed upside down adjacent to their ultimate locations; coated with adhesive or bitumen; and turned over and adhered to the substrate.

Mopping
The application of hot bitumen with a mop or mechanical applicator to the substrate or plies of a bituminous membrane. There are four types of mopping:

- Solid mopping: a continuous coating.
- Spot mopping: bitumen is applied roughly in circular areas, leaving a grid of unmopped perpendicular areas.
- Sprinkle mopping: bitumen is shaken onto the substrate from a broom or mop in a random pattern.
- Strip mopping: bitumen is applied in parallel bands.

Neoprene
A synthetic rubber (polychloroprene) used in liquid and sheet-applied elastomeric roof membranes or flashings.

Nesting
(1) The installation of new metal roof deck directly on top of existing metal roof deck; (2) a method of reroofing with new asphalt shingles over existing shingles in which the top edge of the new shingle is butted against the bottom edge of the existing shingle.

Night seal (or Night Tie-in)
A material and/or method used to temporarily seal a membrane edge during construction to protect the roofing assembly in place from water penetration. Usually removed when roofing application is resumed.

Nitrile alloy
An elastomeric material of synthetic nonvulcanizing polymers.

Nitrile rubber
A membrane whose predominant resinous ingredient is a synthetic rubber made by the polymerization of acrylonitrile with butadiene.

Nondestructive testing (NDT)
A method to evaluate the disposition, strength or composition of materials or systems without damaging the object under test. Typically used to evaluate moisture content in roofing assemblies, the three common test methods are electrical capacitance, infrared thermography and nuclear back-scatter.

Nonwoven Fabric
A textile structure produced by the bonding or interlocking of fibers, or both, accomplished by mechanical, chemical, thermal, or solvent means and combinations thereof.

Nuclear Hydrogen Detection (NHD) Meter
A device that contains a radioactive source to emit high velocity neutrons into a roof system. Reflecting neutrons are measured by a gauge that is used to detect hydrogen; the quantity of hydrogen detected may be linked to the pressure of water.

Nylon
Generic name for a family of polyamide polymers, used as a scrim in some fabric-reinforced sheeting.

Off-ratio Foam
SPF that has excess isocyanate or resin. Off-ratio will not exhibit the full physical properties of normal SPF.

Open Time
The period of time after an adhesive has been applied and allowed to dry, during which an effective bond can be achieved by joining the two surfaces.

Orange Peel Surface Texture
In SPF roofing, a condition of the foam in which the surface shows a fine texture and is compared to the exterior skin of an orange. This surface is considered acceptable for receiving a protective coating.

Overflow Drainage
Component in a roof drainage system used to protect the roof against damage from a water load imposed by blocked or partially blocked primary drainage system; e.g., overflow scupper, overflow interior drain.

Overspray
Undesirable depositions of airborne spray.

Overspray Surface Texture
In SPF roofing, a condition of the foam in which the surface shows a linear coarse textured pattern and/or a pebbled surface. This surface is generally downwind of the sprayed polyurethane path and, if severe, unacceptable for proper coating coverage and protection.

Pan
The bottom flat part of a roofing panel that is between the ribs of the panel.

Partially Attached
A roofing assembly in which the membrane has been "spot affixed" to a substrate, usually with an adhesive or a mechanical device.

Parting Agent
A material applied to one or both surfaces of a sheet to prevent blocking.

Pass
(1) A layer of material, usually applied by the spray method, that is allowed to reach cure before another layer ("pass") is applied; (2) a term used to explain a spray motion of the foam gun in the application of the spray polyurethane foam (SPF) material. The speed of the pass controls the thickness of the SPF.

Pass Line
The junction of two passes of SPF. A distinct line is formed by the top skin of the bottom pass and the next pass adhering to this skin.

Peel Strength
The average load per unit width required to separate progressively a flexible member from a rigid member or another flexible member.

Penetration
(1) Any construction (e.g., pipes, conduits, HVAC supports) passing through the roof; (2) the consistency of a bituminous material expressed as the distance, in tenths of a millimeter (0.1 mm), that a standard needle penetrates vertically into a sample of material under specified conditions of loading, time and temperature.

Perlite
An aggregate used in lightweight insulating concrete and preformed perlitic insulation boards, formed by heating and expanding siliceous volcanic glass.

Phased Application
The installation of a roofing or waterproofing system during two or more separate time intervals or different days. Application of surfacings at different time intervals are typically not considered phased application (*see Surfacing*). A roofing system not installed in a continuous operation.

Picture Framing
A square or rectangular pattern of ridges in a roof membrane or covering over insulation or deck joints.

Pinhole
A tiny hole in a coating, film, foil, membrane or laminate comparable in size to one made by a pin.

Pitch
(*see Coal Tar*)

Pitch-pocket (Pitch-pan)
A flanged, open bottomed enclosure made of sheet metal or other material, placed around a penetration through the roof, filled with grout and bituminous or polymeric sealants to seal the area around the penetration.

Plastic Cement
A roofing industry generic term used to describe asphalt roof cement that is a trowelable mixture of solvent-based bitumen, mineral stabilizers, and other fibers and/or fillers. Generally, intended for use on relatively low slopes, not vertical surfaces. (*see Asphalt Roof Cement and Flashing Cement*)

Plasticizer
A material incorporated in a material to increase its ease of workability, flexibility or distensibility.

Plasticizer Migration
In some thermoplastic roofing membranes, the loss of plasticizer chemicals from the membrane, resulting in shrinkage and embrittlement of the membrane, typically PVC.

Ply
A layer of felt or ply sheet in a built-up roof membrane or roof system.

Polychloroprene
(*see Neoprene*)

Polyester
A polymer in which the repeated structural unit in the chain is of the ester type.

Polyisobutylene (PIB)
A product formed by the polymerization of isobutylene. May be compounded for use as a roof membrane material.

Polymer
A macromolecular material formed by the chemical combination of monomers having either the same or different chemical composition.

Polymer Modified Bitumen
(*see Modified Bitumen*)

Polymeric Methylene Diphenyl Diisocyanate (PMDI)
Component A in SPF. An organic chemical compound having two reactive isocyanate groups. It is mixed with the B component to form polyurethane.

Polymerization
A chemical reaction in which monomers are linked together to form polymers.

Polypropylene
A polymer prepared by the polymerization of propylene as the sole monomers.

Polystyrene
A polymer prepared by the polymerization of styrene as the sole monomer.

Polyvinyl Chloride (PVC)
A synthetic thermoplastic polymer prepared from vinylchloride. PVC can be compounded into flexible and rigid forms through the use of plasticizers, stabilizers, fillers and other modifiers. Rigid forms are used in pipes; flexible forms are used in the manufacture of sheeting and roof membrane materials.

Ponding
The excessive accumulation of water at low-lying areas on a roof that remains after the 48 hours after the end rainfall under conditions conducive to drying.

Popcorn Surface Texture
In SPF roofing, the condition in which the foam surface shows a coarse texture where valleys form sharp angles. This surface is unacceptable for proper coating and protection.

Positive Drainage
The drainage condition in which consideration has been made during design for all loading deflections of the deck and additional roof slope has been provided to ensure drainage of the roof area within 48 hours following rainfall during conditions conducive to drying.

Pourable Sealer
A type of sealant often supplied in two parts and used at difficult-to-flash penetrations, typically in conjunction with pitch-pockets to form a seal.

Primer
(1) A thin, liquid-applied solvent-based bitumen that may be applied to a surface to improve the adhesion of subsequent applications of bitumen; (2) a material that is sometimes used in the process of seaming single-ply membranes to prepare the surfaces and increase the strength (in shear and peel) of the field splice; (3) a thin liquid-applied material that may be applied to the surface of SPVF to improve the adhesion of subsequent application of SPVF protective coatings.

Proportioner
The basic pumping unit for SPF or two-component coating systems. Consists of two positive displacement pumps designed to dispense two components at a precisely controlled ratio.

Protected Membrane Roof (PMR)
An insulated and ballasted roofing assembly in which the insulation and ballast are applied on top of the membrane (sometimes referred to as an "inverted roof assembly").

Protection Mat
A sacrificial material used to shield one roof system component from another.

R-value
(*see Thermal Resistance*)

Re-cover
The addition of a new roof membrane or steep-slope roof covering over a major portion of an existing roof assembly. This process does not involve removal of the existing roofing.

Reinforced Membrane
A roofing or waterproofing membrane that has been strengthened by the addition or incorporation of one or more reinforcing materials, including woven or nonwoven glass fibers, polyester mats or scrims, nylon, or polyethylene sheeting.

Reroofing
The process of re-covering, or tearing-off and replacing an existing roof system.

Resin
Component B in SPF. This component contains a catalyst, blowing agent, fire retardants, surfactants and polyol. It is mixed with the A component to form polyurethane.

Ridging
(*see Buckle*)

Roof Assembly
An assembly of interacting roof components including the roof deck, vapor retarder (if present), insulation and roof covering.

Roof Cement
(*see Asphalt Roof Cement or Coal Tar Roof Cement*)

Roof Covering
The exterior roof cover or skin of the roof assembly, consisting of membrane, panels, sheets, shingles, tiles, etc.

Roof Seamer
(1) Machine that crimps neighboring metal roof panels together; (2) machine that welds laps of membrane sheets together using heat, solvent or dielectric energy.

Roof Slope
The angle a roof surface makes with the horizontal, expressed as a ratio of the units of vertical rise to the units of horizontal length (sometimes referred to as "run"). For English units of measurement, when dimensions are given in inches, slope may be expressed as a ratio of rise to run, such as 4:12 or as an angle.

Roof System
A system of interacting roof components, generally consisting of a membrane or

primary roof covering and roof insulation (not including the roof deck) designed to weatherproof and, sometimes, to improve the building's thermal resistance.

Rosin Paper (specifically Rosin-sized Sheathing Paper)
A nonasphaltic paper used as a sheathing paper or slip sheet in some roof systems.

Rubber
A material that is capable of recovering from large deformations quickly and forcibly.

Saddle
A small tapered/sloped roof area structure that helps to channel surface water to drains. Frequently located in a valley. A saddle is often constructed like a small hip roof or pyramid with a diamond-shaped base. (*see Cricket*)

Sag
Undesirable excessive flow in material after application to a surface.

Scrim
A woven, nonwoven or knitted fabric composed of continuous strands of material used for reinforcing or strengthening membranes.

Scupper
Drainage device in the form of an outlet through a wall, parapet wall or raised roof edge lined with a soldered sheet metal sleeve.

Scuttle
A hatch that provides access to the roof from the interior of the building.

Sealer
A coating designed to prevent excessive absorption of finish coats into porous surfaces; a coating designed to prevent bleeding.

Seam
A joint formed by mating two separate sections of material. Seams can be made or sealed in a variety of ways, including adhesive bonding, hot-air welding, solvent welding, using adhesive tape, sealant, etc.

Seam Sample
In single-ply and sometimes modified bitumen membrane roofing, a sample from the membrane that extends through the side lap of adjacent rolls of membrane, taken for the purpose of assessing the quality of the seam.

Self-adhering Membrane
A membrane that can adhere to a substrate and to itself at overlaps without the use of an additional adhesive. The undersurface of a self-adhering membrane is pro-

tected by a release paper or film, which prevents the membrane from bonding to itself during shipping and handling.

Selvage
(1) An edge or edging that differs from the main part of a fabric, granule-surfaced roll roofing or cap sheet, or other material; (2) a specially defined edge of the material (lined for demarcation), which is designed for some special purpose, such as overlapping or seaming.

Separator Layer
(see Slip Sheet)

Service Temperature Limits
The minimum or maximum temperature at which a coating, SPF or other material will perform satisfactorily.

Shading
Slight differences in surfacing color, such as shingle granule coloring, that may occur as a result of manufacturing operations.

Single-lock Standing Seam
A standing seam that uses one overlapping interlock between two seam panels, in contrast with the double interlocking used in a double standing seam.

Single-ply Membranes
Roofing membranes that are field applied using just one layer of membrane material (either homogeneous or composite) rather than multiple layers.

Single-ply Roofing
A roofing system in which the principal roof covering is a single layer flexible membrane often thermoset or thermoplastic membrane.

Skinning
The formation of a dense film on the surface of a liquid coating or mastic.

Slip Sheet
Sheet material, such as reinforced kraft paper, rosin-sized paper, polyester scrim or polyethylene sheeting, placed between two components of a roof assembly (such as between membrane and insulation or deck) to ensure that no adhesion occurs between them and to prevent possible damage from chemical incompatibility, wearing or abrasion of the membrane.

Slit Sample
In SPF roofing, a small cut about 1 inch x ½ inch x ½ inch [25 mm x 13 mm x 13 mm], in a half-moon shape, used to measure coating film thickness.

Smooth Surface Texture
In SPF roofing, the condition of the foam in which the surface shows spray undulation and is ideal for receiving a protective coating.

Smooth-surfaced Roof
A roof membrane without mineral granule or aggregate surfacing.

Snow Guard
A series of devices attached to the roof in a pattern that attempts to hold snow in place, thus preventing sudden snow or ice slides from the roof; any device intended to prevent snow from sliding off a roof.

Solids Content
The percentage by weight of the nonvolatile matter in an adhesive.

Solvent Cleaners
Used to clean some single-ply roofing membranes prior to splicing, typically including heptane, hexane, white gasoline and unleaded gasoline.

Solvent Welding
A process where a liquid solvent is used to chemically weld or join together two or more layers of certain membrane materials (usually thermoplastic).

SPF Compound
A term used to describe the raw materials (isocyanate and resin) used to make polyurethane foam.

Splice Plate
A metal plate placed underneath the joint between two pieces of metal.

Splice-tape
Cured or uncured synthetic rubber tape used for splicing membrane materials.

Split
A membrane tear resulting from tensile stresses.

Sprayed Polyurethane Foam (SPF)
A foamed plastic material, formed by spraying two components, PMDI (A component) and a resin (B component), to form a rigid, fully adhered, water-resistant and insulating membrane.

Spread Coating
A manufacturing process in which membranes are formed using a liquid compound that is spread onto a supporting reinforcement base layer and then dried to its finished condition.

Spud
To remove the roofing aggregate and most of the bituminous top coating by scraping and chipping.

Square
A unit used in measuring roof area equivalent to 100 square feet [9.29 m^2] of roof area.

Squeegee
(1) A blade of leather or rubber set on a handle and used for spreading, pushing or wiping liquid material on, across or off a surface; (2) to smooth, wipe or treat with a squeegee.

Standing Seam
In metal roofing, a type of seam between adjacent sheets of material made by turning up the edges of two adjacent metal panels and then folding or interlocking them in a variety of ways.

Steep-slope Roofs
A category of roofing that generally includes water-shedding types of roof coverings installed on slopes exceeding 3:12 (14 degrees).

Stiffener Rib
Small intermediate bends in a metal pan used to strengthen the panel.

Strapping (Felts)
A method of installing roofing rolls or sheet good materials parallel with the slope of the roof.

Striations
A parallel series of small grooves, channels or impressions typically within a metal roof panel used to help reduce the potential for oil-canning.

Strippable Films
(For metal) Added protection of plastic films sometimes applied to coated or finished metals after the coil coating process. Applied after prime and top coats to resist damage to the finish prior to and during shipping, fabrication and installation.

Stripping or Strip-flashing
Membrane flashing strips used for sealing or flashing metal flashing flanges into the roof membrane.

Stripping-in
Application of membrane stripping ply or plies.

Structural Panel
A metal roof panel designed to be applied over open framing rather than a continuous or closely spaced roof deck.

Styrene Butadiene Rubber
High molecular weight polymers having rubber-like properties, formed by the random copolymerization of styrene and butadiene monomers.

Styrene Butadiene Styrene Copolymer (SBS)
High molecular weight polymers that have both thermoset and thermoplastic properties, formed by the block copolymerization of styrene and butadiene monomers. These polymers are used as the modifying compound in SBS polymer modified asphalt roofing membranes to impart rubber-like qualities to the asphalt.

Sump
An intentional depression around a roof drain or scupper that promotes drainage.

Sump Pan
A metal pan used to create a depression around a drain or scupper to enhance drainage.

Surface Texture
The resulting surface from the final pass of SPF. The following terms are used to describe the different SPF surface textures: smooth orange peel, coarse orange peel, verge of popcorn, popcorn, treebark and oversprayed.

Surfacing
The top layer or layers of a roof covering specified or designed to protect the underlying roofing from direct exposure to the weather.

Surfactant
Contraction for "surface active agent"; a material that improves the emulsifying, dispersing, spreading, wetting or other surface-modifying properties of liquids.

Tack-free Time
In SPF-based roofing, a curing phase of polyurethane foam when the material is no longer sticky. When the polyurethane foam is tack free, it can be sprayed over with another pass, referred to as a "lift." With some care the polyurethane foam can be walked on soon after it is tack free.

Talc
Whitish powder applied at the factory to the surface of some roofing materials (e.g., vulcanized EPDM membranes); used as a release agent to prevent adhesion of the membrane to itself.

Tapered Edge Strip
A tapered insulation strip used to (1) elevate and slope the roof at the perimeter

and at curbs, and (2) provide a gradual transition from one layer of insulation to another.

Taping
(1) The technique of connecting joints between insulation boards or deck panels with tape; (2) the technique of using self-adhering tape-like materials to seam or splice single-ply membranes.

Tar
A brown or black bituminous material, liquid or semi-solid in consistency, in which the predominating constituents are bitumens obtained as condensates in the processing of coal, petroleum, oil-shale, wood or other organic materials.

Tear-off and Reroof
The removal of all roof system components down to the structural deck, followed by installation of a completely new roof system.

Tear Resistance
The load required to tear a material, when the stress is concentrated on a small area of the material by the introduction of a prescribed flaw or notch. Expressed in psi (pounds force) per inch width or kN/m (kilonewton per meter width).

Tear Strength
The maximum force required to tear a specimen.

Tension Leveling
The process of pulling metal coil stock between two spools under a certain pressure to help reduce side camber and potential oil canning in the coil stock caused by manufacturing and cutting processes.

Test Cut
A sample of the roof system or assembly which exposes the roof deck and is used to diagnose the condition of the membrane, or evaluate the type and number of plies or number of membranes, or rates of application (e.g., the weight of the average interply bitumen moppings).

Thermal Resistance (R)
Under steady conditions, thermal resistance is the mean temperature difference between two defined surfaces of material or construction that induces unit heat flow through a unit area. In English (inch·pound) units it is expressed as F ft^2 h/Btu.

Note 1: A thermal resistance (R) value applies to a specific thickness of a material or construction.

Note 2: The thermal resistance (R) of a material is the reciprocal of the thermal conductance (C) of the same material (i.e., R = 1/C).

Note 3: Thermal resistance (R) values can be added, subtracted, multiplied and divided by mathematically appropriate methods.

Thermal Shock
The stress-producing phenomenon resulting from sudden temperature changes in a roof membrane when, for example, a cold rain shower follows brilliant sunshine.

Thermal Stress
Stress introduced by uniform or non-uniform temperature change in a structure or material that is contained against expansion or contraction.

Thermal Transmittance (U or U-factor)
Thermal transmittance (U or U-factor) is the time rate of heat flow per unit area under steady conditions from the fluid (e.g., air) on the warm side of a barrier to the fluid (e.g., air) on the cold side, per unit temperature difference between the fluids. In English (inch·pound) units expressed as Btu/h ft^2F.

Note 1: A thermal transmittance (U) value applies to the overall thermal performance of a system (e.g., roof assembly).

Note 2: Thermal transmittance (U) is sometimes called the overall coefficient of heat transfer.

Note 3: Thermal transmittance (U) is reciprocal of the overall thermal resistance (RT) of a system (i.e., U = 1/RT).

Thermoplastic
A material that softens when heated and hardens when cooled. This process can be repeated provided that the material is not heated above the point at which decomposition occurs.

Thermoplastic Olefin Membrane (TPO)
A blend of polypropylene and ethylene-propylene polymers. Colorant, flame retardants, UV absorbers and other proprietary substances which may be blended with the TPO to achieve the desired physical properties. The membrane may or may not be reinforced.

Thermoset
A class of polymers that, when cured using heat, chemical or other means, changes into a substantially infusible and insoluble material.

Tie-in
In roofing and waterproofing, the transitional seal used to terminate a roofing or waterproofing application at the top or bottom of flashings or by forming a watertight seal with the substrate, membrane, or adjacent roofing or waterproofing system.

T-joint
The condition created by the overlapping intersection of three or four sheets in the membrane.

Torch-applied
Method used in the installation of polymer modified bitumen membranes characterized by using open flame propane torch equipment.

Transverse Seam
The joint between the top of one metal roof panel and the bottom of the next panel, which runs perpendicular to the roof slope.

Treebark Surface Texture
In SPF roofing, the surface condition of the foam which shows a coarse texture where valleys form sharp angles. This surface is unacceptable for proper coating and protection.

U-Value
(*see Thermal Transmittance*)

Underlayment
An asphalt-saturated felt or other sheet material (may be self-adhering) installed between the roof deck and roof covering, usually used in steep-slope roof construction. Underlayment is primarily used to separate the roof covering from the roof deck, shed water and provide secondary weather protection for the roof area of the building.

Vapor Retarder
A layer(s) of material or a laminate used to appreciably reduce the flow of water vapor into a roof assembly.

Verge of Popcorn Texture
In SPF roofing, the verge of popcorn surface texture is the roughest texture suitable for receiving the protective coating on a sprayed polyurethane foam roof. The surface shows a texture where nodules are larger than valleys, with the valleys relatively cured. This surface is acceptable for receiving a protective coating only because of the relatively cured valleys. However, the surface is considered undesirable because of the additional amount of coating material required to protect the surface properly.

Vermiculite
An aggregate used in lightweight insulating concrete, formed by heating and expanding of a micaceous material.

Volatile Organic Compounds (VOC)
Means any compound of carbon, excluding carbon monoxide, carbon dioxide, carbonic acid, metallic carbides or carbonates, and ammonium carbonate, which participate in atmospheric photochemical reactions.

Vulcanization
An irreversible process during which a rubber compound, through a change in its chemical structure (for example, cross-linking), becomes less plastic and more resistant to swelling by organic liquids and elastic properties are conferred, improved, or extended over a greater range of temperature.

Warm Roof Assembly
A roof assembly configured with each component placed immediately on top of the preceding component; each component is in contact with the adjacent component. No space is provided for ventilation of the roof assembly. Also known as a "compact" roof assembly.

Wash Coat
A primer typically provided on the back side of painted metal products to help protect the underlying metal from wear and corrosion.

Water-shedding
The ability of individual, overlapping components to resist the passage of water without hydrostatic pressure.

Wet Film Thickness
The thickness, expressed in mils, of a coating or mastic as applied but not cured. *(For comparison, see Dry Film Thickness.)*

Wind Uplift
The force caused by the deflection of wind at roof edges, roof peaks or obstructions, causing a drop in air pressure immediately above the roof surface.

Yield
In SPF-based roofing, the volume of foam per unit weight, normally expressed as board feet per pound or board feet per 1,000 pounds.

American Institute of Architects. 1997. *Buildings at Risk: Wind Design Basics for Practicing Architects.* Washington, DC: AIA.

American Society of Civil Engineers. 2000. *Minimum Design Loads for Buildings and Other Structures.* ASCE 7-98. Reston, VA: ASCE.

ARCOM. *MASTERSPEC™.* A product of the AIA. Salt Lake City, UT: ARCOM (updated quarterly).

Asphalt Roofing Manufacturers Association. 1997. *Modified Bitumen Design Guide for Building Owners.* 430-MBS-97. Calverton, MD: ARMA.

Asphalt Roofing Manufacturers Association and National Roofing Contractors Association. 1993. *Quality Control Guidelines for the Application of Built-Up Roofing.* Rosemont, IL: NRCA.

———. 1996a. *Manual for Inspection and Maintenance of Built-Up and Modified Bitumen Roof Systems: A Guide For Building Owners.* Rosemont, IL: NRCA.

———. 1996b. *Quality Control Guidelines for the Application of Polymer Modified Bitumen Roofing.* Rosemont, IL: NRCA.

Boivin, Marie-Anne. 1997. "Extensive Green Roofs." In *Proceedings of the Fourth International Symposium on Roofing Technology.* Rosemont, IL: NRCA.

Burch, Douglas M. 1986. "A Heat Transfer Analysis of Metal Fasteners in Low-slope Roofs." *Roofing Research Standards Development.* ASTM STP 959. Philadelphia: ASTM.

Chaize, A. 1995. La retenue temporaire des eaux pluviales: une fonction economique et ecologique des toitures a faible pente. In *Proceedings of the International Waterproofing Association IX International Congress.* Nottingham, UK: International Waterproofing Association.

CIB/RILEM Joint Committee on Roofing Materials and Systems. 2001a. *Condition Assessment of Roofs.* Online document. Available on web site (www.cibworld.nl).

———. 2001b. *Towards Sustainable Roofing.* Online document. Available on web site (www.cibworld.nl).

Desjarlais, A.O. 1995. "Self-drying Roofs: What! No Dripping!" In *Proceedings of the ASHRAE/DOE/BTECC Conference.* Atlanta: ASHRAE.

Desjarlais, A.O., and A.N. Karagiozis. 1999. "Review of Existing Criteria and Proposed Calculations for Determining the Need for Vapor Retarders." In *Proceedings of the North American Conference on Roofing Technology.* Rosemont, IL: NRCA.

BIBLIOGRAPHY

Dupuis, Rene M. 1998. *A Field and Laboratory Assessment of Sprayed Polyurethane Foam-Based Roof Systems.* Rosemont, IL: National Roofing Foundation.

Factory Mutual Global. *Loss Prevention Data for Roofing Contractors.* Norwood, MA: FMG (dates vary).

Factory Mutual Research. *Approval Guide.* Norwood, MA: FMR (updated periodically).

———. *Specifications Tested Products Guide.* Norwood, MA: FMR (updated quarterly).

Federal Emergency Management Agency and Government of Guam. 1998. *Typhoon Paka: Observations and Recommendations on Building Performance and Electrical Power Distribution System, Guam, U.S.A.* San Francisco: FEMA.

Griffin, C.W., and R.L. Fricklas. 1996. *The Manual of Low-Slope Roof Systems.* 3rd ed. New York: McGraw-Hill.

Laaly, H.O. 1992. *The Science and Technology of Traditional and Modern Roofing Systems.* Los Angeles: Laaly Scientific Publishing.

Metal Building Manufacturers Association. 2000. *Metal Roofing Systems Design Manual.* 1st ed. Cleveland, OH: MBMA.

Midwest Roofing Contractors Association and National Roofing Contractors Association. 1996. *APP and SBS Modified Bitumen Membrane Roofs: A Survey of Field Performance.* Rosemont, IL: NRCA.

Mock, Luther C. 2000. "Roof Design by Weighted Evaluation." In *Proceedings of the 15th International Convention & Trade Show.* Raleigh, NC: Roof Consultants Institute.

National Council of Architectural Registration Boards. 2000. *Low-Slope Roofing I,* by Justin Henshell, Donald Brotherson and Paul Buccellato. Washington, DC: NCARB.

———. 2001. *Energy-Conscious Architecture,* by Carl Stein. Washington, DC: NCARB.

———. 2001. *Sustainable Design,* by Jonee Kulman and Joel Schurke. Washington, DC: NCARB.

National Roofing Contractors Association. 1999. *Considerations Pertaining to Polyisocyanurate Insulation.* Special report. Rosemont, IL: NRCA.

———. 2000. *Use of Cover Boards over Polyisocyanurate Insulation.* Bulletin 2000-3. Rosemont, IL: NRCA.

———. 2001a. *Low-Slope Roofing Materials Guide.* Rosemont, IL: NRCA (published every other year).

———. 2001b. *The NRCA Roofing and Waterproofing Manual.* 5th ed. 3 volumes. Rosemont, IL: NRCA.

National Roofing Contractors Association and The Society of the Plastics Industry/Spray Polyurethane Foam Division. 1998a. *Manual for Inspection and Maintenance of Spray Polyurethane Foam-Based Roof Systems: A Guide for Building Owners.* Rosemont, IL: NRCA.

———. 1998b. *Quality Control Guidelines for the Application of Sprayed Polyurethane Foam Roofing.* Rosemont, IL: NRCA.

National Roofing Contractors Association and SPRI. 1997. *Quality Control Guidelines for the Application of Thermoset Single-Ply Roof Membranes.* Rosemont, IL: NRCA.

Oak Ridge National Laboratory. 1994. *Proceedings of the Low-Slope Reroofing Workshop.* CONF-9405206. Oak Ridge, TN: ORNL.

———. 1997. *Proceedings of the Sustainable Low-Slope Roofing Workshop.* CONF-9610200. Oak Ridge, TN: ORNL.

Paroli, Ralph, Ana Delgado, Om Dutt, Thomas L. Smith, and Terrence Simmons. 1995. "Shrinkage of EPDM Roof Membranes: Phenomenon, Causes, Prevention and Remediation." In *Proceedings of the 11th Conference on Roofing Technology.* Rosemont, IL: NRCA.

Robinson, Jack, Dane Bradford, Jeffrey Mitchell, and Paul Majkowski. 1999. "A Comparison of Three Different Technologies for Performing Nondestructive Roof Moisture Survey." In *Proceedings of the North American Conference on Roofing Technology.* Rosemont, IL: NRCA.

Rossiter, Walter, and R.G. Mathy. 1986. *A Methodology for Assessing the Thermal Performance of Low-Sloped Roofing Systems.* NBSIR 85-3264. Gaithersburg, MD: National Institute of Standards and Technology.

Schmeltz, Rachel S., and Timothy M. Davey. 1999. "Energy Star® Label for Roofing Products, and a Contractor's View of the Energy Star® Label for Roofing Products." In *Proceedings of the North American Conference on Roofing Technology.* Rosemont, IL: NRCA.

Sheet Metal and Air Conditioning Contractors' National Association, Inc. 1993. *Architectural Sheet Metal Manual.* 5th ed. Chantilly, VA: SMACNA.

Smith, Thomas L. 1993. "How Did PUF Roofs Perform During Hurricane Andrew?" *Professional Roofing 23* (1): 20-25 (January).

———. 1995a. "Improving Wind Performance of Mechanically attached Single-ply Membrane Roof Systems: Lessons from Hurricane Andrew." In *Proceedings of the International Waterproofing Association IX International Congress.* Nottingham, UK: International Waterproofing Association.

———. 1995b. "Insights on Metal Roof Performance in High-wind Regions." *Professional Roofing.* 25 (2): 12-16 (February).

———. 1995c. "Deck Attachment: Anchoring the Roof Covering's Foundation." *Professional Roofing.* 25 (12): 50 (December).

———. 1997. "Sustainable Roofing: A Paradigm of Challenges and Opportunities." In *Proceedings of the Sustainable Low-Slope Roofing Workshop.* Oak Ridge, TN: ORNL.

———. 2000. "The Roof System Blew Off—Now What?" *Professional Roofing.* 30 (8): 28-31 (August).

———. 2001a. "Benefits of Reflective Roofs in Northern Areas of the U.S." *RCI Interface.* XIX (8): 4-12 (August).

———. 2001b. "Designing for Future Reroofing: Recommendations for Enhanced Sustainability." In *Proceedings of the International Conference on Building Envelope Systems and Technologies (ICBEST), Volume 2.* Ottawa: National Research Council of Canada.

———. 2001c. "Uplift Resistance of Existing Roof Decks: Recommendations for Enhanced Attachment During Reroofing Work." In *Proceedings of the International Conference on Building Envelope Systems and Technologies (ICBEST), Volume 1.* Ottawa: National Research Council of Canada.

Smith, Thomas L., and Dr. James R. McDonald. 1993. "Preliminary Design Guidelines for Wind-resistant Roofs on Essential Facilities." In *Proceedings of the 7th U.S. National Conference on Wind Engineering.* Los Angeles: The Wind Engineering Research Council.

Smith, Thomas L., and Rene M. Dupuis. 1996. "Sprayed Polyurethane Foam Roofing: Overcoming Aesthetic Barriers." In *Proceedings of the Polyurethanes Expo '96.* Washington, DC: Polyurethane Division of The Society of the Plastics Industry, Inc.

Smith, Thomas L., Dr. J. Kind, and Dr. R.J. McDonald. 1992. "Hurricane Hugo: Evaluation of Wind Performance and Wind Design Guidelines for Aggregate Ballasted Single-ply Membrane Roof Systems." In *Proceedings Acociacion Internacional De La Impermeabilizacion VIII Congreso Internacional.* Nottingham, UK: International Waterproofing Association.

SPRI. 1997. *Wind Design Standard for Ballasted Single-ply Roofing Systems.* ANSI/SPRI RP-4. Needham, MA: SPRI.

———. 1998. *Wind Design Standard for Edge Systems Used with Low Slope Roofing Systems.* ANSI/SPRI ES-1. Needham, MA: SPRI.

———. 2001a. *Flexible Membrane Roofing: A Professional's Guide to Specifications.* Needham, MA: SPRI.

———. 2001b. *Standard Field Test Procedure for Determining the Withdrawal Resistance of Roofing Fastener.* ANSI/SPRI FX-1. Needham, MA: SPRI.

SPRI and National Roofing Contractors Association. 1992. *Manual of Roof Inspection, Maintenance, and Emergency Repair for Existing Single-Ply Roofing Systems.* Rosemont, IL: NRCA.

Underwriters Laboratories Inc. *Fire Resistance Directory Volume I.* Northbrook, IL: UL (updated annually).

———. *Roofing Materials and Systems Directory.* Northbrook, IL: UL (updated annually).

———. *Roof Deck Constructions.* Northbrook, IL: UL (updated annually).

ORGANIZATIONS

American Association for Wind Engineering (AAWE)
web site: www.aawe.org

The American Institute of Architects (AIA)
1735 New York Avenue, NW
Washington, DC 20006-5292
telephone: (202) 626-7300
fax: (202) 626-7420
web site: www.aiaonline.com

American Society for Testing and Materials (ASTM)
100 Barr Harbor Drive
West Conshohocken, PA 19428-2959
telephone: (610) 832-9585
fax: (610) 832-9555
web site: www.astm.org

American Society of Civil Engineers (ASCE)
1801 Alexander Bell Drive
Reston, VA 20191-4400
telephone: (800) 548-2723
fax: (703) 295-6333
web site: www.asce.org

Asphalt Roofing Manufacturers Association (ARMA)
1156 15th Street NW, Suite 900
Washington, DC 20005
telephone: (202) 207-0917
fax: (202) 223-9741
web site: www.asphaltroofing.org

Cold Regions Research and Engineering Laboratory (CRREL)
U.S. Army Corps of Engineers
72 Lyme Road
Hanover, NH 03755-1290
telephone: (603) 646-4588
fax: (603) 646-4640
web site: www.crrel.usace.army.mil/

Energy Star Programs
U.S. Environmental Protection Agency
Mail Code 6202J
401 M Street, SW
Washington, DC 20460
telephone: (888) 782-7937
fax: (202) 775-6680
web site: www.energystar.gov

Factory Mutual Research (FMR)
Northeast Field Engineering
P.O. Box 9102
Norwood, MA 02062
telephone: (781) 762-8000
fax: (781) 762-8718
web site: www.fmglobal.com

Federal Emergency Management Agency (FEMA)
500 C Street, SW
Washington, DC 20472
telephone: (202) 646-4600
fax: (202) 646-2577
web site: www.fema.gov/mit

International Council for Research and Innovation in Building and Construction (CIB)
Postbox 1837
3000 BV Rotterdam, The Netherlands
telephone: 31 10 411 0240
fax: 31 10 433 4372
web site: www.cibworld.nl

International Union of Testing and Research Laboratories for Materials and Structures (RILEM)
ENS – Bâtiment Cournot, 61 avenue du Président Wilson
F-94235 Cachan Cedex, France
telephone: 33 14 740 2397
fax: 33 14 740 0113
web site: www.rilem.org

Metal Building Manufacturers Association (MBMA)
1300 Sumner Avenue
Cleveland, OH 44115-2851
telephone: (216) 241-7333
fax: (216) 241-0105
web site: www.mbma.com

Metal Construction Association (MCA)
4700 W. Lake Avenue
Glenview, IL 60025
telephone: (847) 375-3675
fax: (847) 665-2234
web site: www.metalconstruction.org

National Roofing Contractors Association (NRCA)
10255 W. Higgins Road, Suite 600
Rosemont, IL 60018
telephone: (847) 299-9070
fax: (847) 299-1183
web site: www.nrca.net

Oak Ridge National Laboratory (ORNL)
P.O. Box 2008, MS-6070
Oak Ridge, TN 37831-6070
telephone: (865) 574-5206
fax: (865) 574-5227
web site: www.ornl.gov/btc

Roof Coatings Manufacturers Association (RCMA)
1156 15th Street, NW, Suite 900
Washington, DC 20005
telephone: (202) 207-0919
fax: (202) 223-9741
web site: www.roofcoatings.org

Roof Consultants Institute (RCI)
7424 Chapel Hill Road
Raleigh, NC 27607-5041
telephone: (919) 859-0742
fax: (919) 859-1328
web site: www.rci-online.org

Roofing Industry Educational Institute (RIEI)
10255 W. Higgins Road, Suite 600
Rosemont, IL 60018
telephone: (847) 299-9070
fax: (847) 299-1183
web site: www.riei.org

Sheet Metal and Air Conditioning Contractors' National Association, Inc. (SMACNA)
4201 Lafayette Center Drive
Chantilly, VA 20151-1209
telephone: (703) 803-2980
fax: (703) 803-3732
web site: www.smacna.org

Spray Polyurethane Foam Alliance (SPFA)
(formerly the Sprayed Polyurethane Foam Division, SPFD, of The Society of the Plastics Industry, Inc.)
1300 Wilson Boulevard, Suite 800
Arlington, VA 22209
telephone: (800) 523-6154
fax: (703) 253-0664
web site: www.sprayfoam.org

SPRI
230 Reservoir Street, Suite 309A
Needham, MA 02494
telephone: (781) 444-0242
fax: (781) 444-6111
web site: www.spri.org

Underwriters Laboratories Inc. (UL)
333 Pfingsten Road
Northbrook, IL 60062-2096
telephone: (847) 272-8800
fax: (847) 272-2020
web site: www.ul.com

PERIODICALS

Professional Roofing
NRCA
10255 W. Higgins Road, Suite 600
Rosemont, IL 60018
telephone: (847) 299-9070
fax: (847) 299-1183
web site: www.nrca.net

RCI Interface
RCI
7274 Chapel Hill Road
Raleigh, NC 27607-5041
telephone: (919) 859-0742
fax: (919) 859-1328
web site: www.rci-online.org

Roofing Contractor
Business News Publishing Company
P.O. Box 7021
Troy, MI 48007-7021
telephone: (248) 362-3700
fax: (248) 362-5103
web site: www.roofingcontractor.com

RSI Roofing/Siding/Insulation
Advanstar Communications, Inc.
131 West First Street
Duluth, MN 55802-2065
telephone: (888) 527-7008
fax: (218) 723-9477
web site: www.rsimag.com

Western Roofing Insulation and Siding
Dodson Publications, Inc.
546 Court Street
Reno, NV 89501-1711
telephone: (775) 333-1080
fax: (775) 333-1081
web site: www.westernroofing.net

INDEX

INDEX

INDEX

N C A R B

PROFESSIONAL DEVELOPMENT PROGRAM

LOW-SLOPE ROOFING II QUIZ

LOW-SLOPE ROOFING II QUIZ

INSTRUCTIONS

Enclosed with this monograph are an answer sheet and a return envelope for your use in completing the following quiz.

When answering the questions, you may refer to the monograph or other reference materials as necessary; however, you must answer the questions without the assistance of others. NCARB will not give advice or clarification on questions.

Upon completion of the quiz, fold the answer sheet where indicated, and return it to NCARB in the supplied envelope, making sure to affix proper postage.

Within 10 working days from receipt at the Council office, you will receive a pass/fail score report. Failing score reports will be accompanied by a new answer sheet and envelope so that you may retake the quiz. NCARB recommends that you undertake additional study of the monograph before retaking the quiz.

Because this quiz is available online, we encourage you to take advantage of the ease and accessibility of completing it and then downloading your acknowledgment via the Internet. The enclosed flier provides additional details about NCARB's online monograph quizzes.

Please retain the score reports for your records and for submission to those state boards requiring continuing professional development for license maintenance. Duplicate copies of your report are available upon written request.

PDP quiz scores will be maintained for five years by NCARB and will be made available only to participants and registration boards.

QUESTION 1

Storm damage research indicates that in high wind conditions, the best performing roof systems are

1. liquid applied.
2. four-ply built-up.
3. sprayed polyurethane foam.
4. torched-on modified bitumen.

A. 1 and 3
B. 2 and 4
C. 3 and 4
D. 1 and 2

QUESTION 2

For a 30-story office building in a downtown business district, the best selection for a roofing system would most likely be

A. adhered single-ply membrane.
B. ballasted single-ply membrane.
C. aggregate-surfaced built-up asphalt.
D. aggregate-surfaced built-up coal tar.

QUESTION 3

Re-covering is defined as the removal of the existing roof system down to the deck.

A. True
B. False

QUESTION 4

The deck with the highest tolerance to moisture is

A. cementitious wood-fiber.
B. galvanized metal.
C. painted metal.
D. gypsum.

QUESTION 5

NRCA Bulletin 2000-03 recommends placing a cover board over polyisocyanurate insulation for ballasted systems to

A. prevent gassing transfer.
B. secure a higher U-factor.
C. provide greater adhesive.
D. protect the foam insulation.

QUESTION 6

The advantages of sprayed polyurethane foam roof include

1. no joints.
2. no mechanical fasteners.
3. greater UV resistance.
4. cool-weather application.

A. 1 and 2
B. 2 and 3
C. 3 and 4
D. 1 and 4

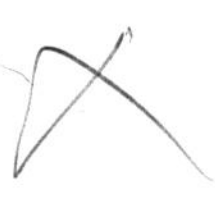

QUESTION 7

When a non-catastrophic premature failure of a roof occurs, the architect of record should first inform the owner to

1. contact his or her lawyer.
2. authorize permanent emergency repairs.
3. notify the contractor and the materials manufacturer.
4. commission an investigation and evaluation of the problem.

A. 1 and 2
B. 2 and 3
C. 3 and 4
D. 1 and 4

QUESTION 8

A long-term warranty might cause the owner to make which of the following incorrect assumptions?

1. There will be no additional cost for the warranty
2. Other insurance coverage is unnecessary
3. Periodic maintenance will be less critical
4. The quality of installation is not critical

A. 1, 2 and 4 only
B. 1, 2 and 3 only
C. 2, 3 and 4 only
D. 1, 3 and 4 only

QUESTION 9

Approving in the field changes from the contract documents may affect which of the following?

1. Code compliance of the roof
2. Area required to lift materials
3. Performance of the roof system
4. Number of roofing subcontractors

A. 1 and 3 only
B. 2 and 4 only
C. 3 and 4 only
D. 1 and 2 only

QUESTION 10

Which of the following should the architect analyze when insulation applied with ribbon foam-applied adhesive delaminates from the deck?

1. Recent wind velocity records
2. The gauge or thickness of the deck
3. The thickness of the insulation material
4. Spacing distance of the ribbon adhesive

A. 1 and 2 only
B. 2 and 3 only
C. 3 and 4 only
D. 1 and 4 only

QUESTION 11

To which of the following roof systems should an architect give primary consideration in areas of frequent rainfall, cool weather and moderate winds?

A. Single-ply
B. Metal panel
C. Foam-applied
D. Liquid-applied

QUESTION 12

What first steps should an architect perform after a premature roof failure?

1. Assign liability to the roofing contractor
2. Prepare a report with video documentation
3. Document conditions with quality photography
4. Give his or her opinion and analysis of the problems

A. 1 and 2 only
B. 1 and 4 only
C. 3 and 4 only
D. 2 and 3 only

QUESTION 13

A written warranty is always an effective solution to dealing with the financial responsibility of premature roof failure.

A. True
B. False

QUESTION 14

A warranty defines the

A. time duration of roof repairs.
B. scope of routine maintenance.
C. amount of insurance coverages to be provided.
D. specific legal rights and obligations of the parties.

QUESTION 15

Which of the following warranty provisions would be considered unfavorable to the owner?

1. Length of coverage is limited to 10 years or less
2. Clauses that stipulate rapid resolution to problems
3. Leak repairs are limited to patching the membrane
4. Manufacturer determines the applicability of the warranty

A. 1 and 3 only
B. 2 and 4 only
C. 3 and 4 only
D. 1 and 2 only

QUESTION 16

An aggregate-surface protected membrane roof (PMR) is limited to which of the following slopes?

A. ½:12 (4 percent)
B. 1:12 (8 percent)
C. 1½:12 (12 percent)
D. 2:12 (16 percent)

LOW-SLOPE ROOFING II QUIZ

QUESTION 17

In order to obtain an acceptable roofing contractor, the architect should specify

A. a roofing certificate.
B. a labor-only warranty.
C. a manufacturer's warranty.
D. qualification requirements.

QUESTION 18

The length of coverage for a contractor's warranty and a manufacturer's warranty are generally not the same.

A. True
B. False

QUESTION 19

Where does the highest roof uplift occur in the diagrammatic building shown below?

A. A
B. B
C. C
D. D

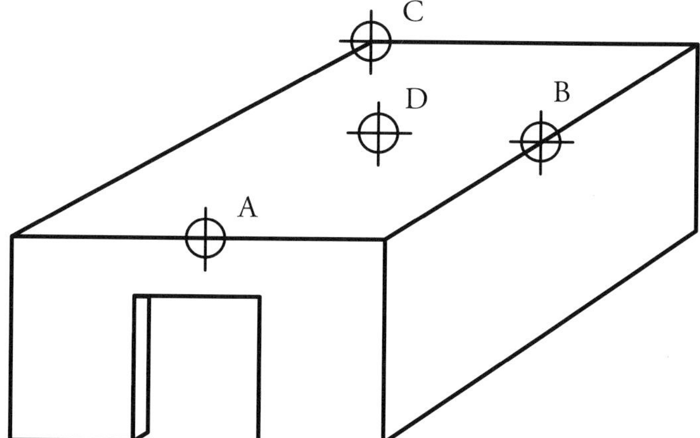

QUESTION 20

When using insulation boards as part of a fully adhered low-slope roof system, the maximum recommended thickness and size for an individual board are

A. 1" ; 4' X 8'.
B. 2" ; 4' X 4'.
C. 2" ; 4' X 8'.
D. 4" ; 4' X 4'.

QUESTION 21

Mechanically attached single-ply roofs are most suitable over lightweight insulating concrete.

A. True
B. False

QUESTION 22

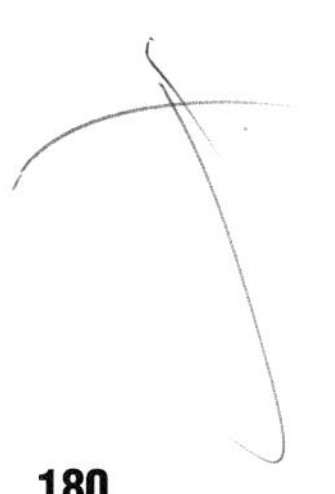

The single design aspect that can significantly affect the performance of a roof system is the

A. slope of the roof.
B. number of seams.
C. material selection.
D. method of fastening.

QUESTION 23

Roof membrane blow-off is almost always a result of the

A. release of fasteners through insulation.
B. lifting and peeling of metal edge flashing.
C. tearing of the membrane due to high winds.
D. application technique of the roofing contractor.

QUESTION 24

In applying a safety factor of 2 in the design of a roof assembly where the design upload lift is 60 psf, what is the Factory Mutual (FM) rating?

A. FM 2-60
B. FM 1-30
C. FM 1-120
D. FM 2-120

QUESTION 25

The fire rating of a roof system can be downgraded if substantially more insulation is specified than that allowed in the tested assembly.

A. True
B. False

QUESTION 26

Sprayed polyurethane foam roofs are best used in densely urbanized areas.

A. True
B. False

QUESTION 27

Which of the following design considerations are most likely to influence the longevity of a low-slope roof for a museum in the western plains of the United States?

A. Aesthetics, roof-top traffic
B. Hail performance, moisture control
C. Building codes, wind performance
D. Energy efficiency, roof system weight

QUESTION 28

Factory Mutual Research provides testing services and standards for roofs in which of the following areas?

A. Foot traffic, color retention, hail-resistance and water leakage
B. Hail-resistance, wind-resistance, fire-resistance and fasteners
C. Wind-resistance, fire-resistance, foot traffic and corrosion-resistance
D. Load characteristics, color retention, corrosion-resistance and building codes

LOW-SLOPE ROOFING II QUIZ

QUESTION 29

A contractor submitted four materials/installation options to prevent long membrane tears. Which of the following options should the architect approve?

A. Air-pressure equalized system
B. Reinforced membrane system
C. Fasteners through the roof insulation
D. Spacing between tab rows at 15 feet on center

QUESTION 30

An APP modified bitumen sheet can be torched, adhered in cold adhesive, or adhered in hot asphalt.

A. True
B. False

QUESTION 31

Which applications are likely to be used when specifying an APP MB base sheet?

A. Hot asphalt adhered and loose laid
B. Heat welded and hot asphalt adhered
C. Cold adhesive adhered and loose laid
D. Cold adhesive adhered and mechanical fasteners

QUESTION 32

SIS modified bitumen sheets are attached by

A. hot asphalt.
B. self-adhesion.
C. cold adhesive.
D. mechanical fasteners.

QUESTION 33

The thermoplastic family of single-ply membranes can be hot-air welded.

A. True
B. False

QUESTION 34

Although PVC membranes contain chlorine and become brittle when in contact with some materials, they offer what advantage compared to TPO?

A. Less expensive
B. Highly reflective
C. More impact-resistant
D. Have a longer track record

QUESTION 35

When specifying a ballasted thermoplastic membrane, which would be a good choice?

A. PIB
B. PVC
C. KEE
D. TPO

QUESTION 36

EPDM sheets are seamed by

1. hot asphalt.
2. hot-air welding.
3. a liquid-applied adhesive.
4. a specially formulated tape.

A. 1 and 3 only
B. 2 and 4 only
C. 3 and 4 only
D. 1 and 2 only

QUESTION 37

When specifying EPDM in a mechanically attached or loose-laid air-pressure equalized configuration, it is prudent to specify a reinforced sheet.

A. True
B. False

QUESTION 38

When specifying a non-reinforced EPDM sheet, other than a fully adhered configuration, the most important factor to pay attention to is the

A. thickness.
B. base securement.
C. puncture resistance.
D. elongation property.

QUESTION 39

Although some metal roofing manufacturers say their systems are suitable for slopes as low as ¼ inch per foot (2 percent), NRCA recommends the following minimum slope of

A. 2:12 (17 percent).
B. 3:12 (25 percent).
C. ½ inch per foot (4 percent).
D. the manufacturer's recommendation.

QUESTION 40

Thermoset materials normally cross-link during manufacturing.

A. True
B. False

QUESTION 41

Which type of installation method is the most reliable for connecting steel deck to structure?

A. Welding
B. Screw-attachment
C. Welding with washers
D. Shot-driven fastener attachment

QUESTION 42

Rather than show flashing and penetration details on the contract drawings, it is typically sufficient to specify that the contractor follow the manufacturer's standard details for long-term reliable service.

A. True
B. False

LOW-SLOPE ROOFING II QUIZ

QUESTION 43

A building near a ski resort has a mineral-surfaced cap sheet BUR on an upper roof that has a 4:12 (33 percent) slope. Below the eave of the upper roof there is a lower roof with a slope of 3:12 (25 percent). What would be suitable replacements for both roofs?

1. PMR
2. EPDM or PVC
3. Metal panels with snow guards
4. Granular-surfaced modified bitumen

A. 1 and 3 only
B. 2 and 4 only
C. 3 and 4 only
D. 1 and 2 only

QUESTION 44

You are designing a new school in a hot humid climate for a district that has a record of poorly funding its maintenance and operations budget. What is the most appropriate roof for this building?

A. White PVC
B. Black EPDM
C. White EPDM
D. APP with coating

QUESTION 45

Liquid-applied membranes have which of the following attributes?

A. Versatility
B. Durability
C. Wind resistance
D. Freeze-thaw resistance

QUESTION 46

Sustainable roof principles can be implemented by designing

1. serviceability and reliability into the roof system.
2. low embodied-energy products into the roof system.
3. materials that can be reused or recycled at the end of the roof's life.
4. a thermally efficient system that considers R-value and reflectivity.

A. 1 and 3
B. 2 and 4
C. 3 and 4
D. 1 and 2

QUESTION 47

After six years of service, the roof membrane experienced a significant premature aging failure, which resulted in a great amount of interior water damage. Which would likely provide the most relief to the building owner?

A. A contractor's warranty
B. A manufacturer's warranty
C. A manufacturer's material-only warranty
D. The Uniform Commercial Code provisions

QUESTION 48

A good master guide specification is preferable to a manufacturer's guide specification because it

1. is more general.
2. has been peer reviewed.
3. is limited for specific projects.
4. has been coordinated with other sections.

A. 1 and 3
B. 3 and 4
C. 1 and 2
D. 2 and 4

QUESTION 49

What measures should the architect take when the building owner does not want to pay the architect for construction administration services relative to a reroofing project?

A. Decline to do the work
B. Allow for alternative roofing systems
C. Require the contractor to submit substitutions directly to the owner
D. Specify a roofing system that is relatively forgiving during application

QUESTION 50

An SBS modified bitumen membrane adhered in cold adhesive was specified by the architect. Due to construction delays, the roof will be installed during cool, damp weather. The contractor requests authorization to torch-apply the membrane at no change in cost. The architect should

A. accept the change, but require the contractor to use the manufacturer's sheets that are intended for torch application.
B. not accept the change; although the substrate is non-combustible, torches on the roof are a fire hazard.
C. in lieu of the proposed change, require the contractor to install self-adhering modified bitumen sheets.
D. in lieu of the proposed change, require the contractor to adhere the sheets in hot asphalt.

QUESTION 51

What is the least common cause of roof failures?

A. Hail damage
B. Wind damage
C. Premature membrane degradation
D. Intermittent, modest or moderate leaks

QUESTION 52

Typically, after completing a wind blow-off investigation, the architect should immediately recommend

A. a temporary roof.
B. a redesign of the roof.
C. the owner sue the contractor.
D. replacement of the roof as specified.

QUESTION 53

In NRCA's *1999-2000 Annual Market Survey,* material other than BUR amounted to 19.5 percent of new construction and 21.3 percent of the reroofing in the 1999 U.S. low-slope market.

A. True
B. False

QUESTION 54

Which is the proper order for the four parts of the first phase in the low-slope design process?

1. Understand membrane materials
2. Respond to project requirements
3. Prepare specifications and drawings
4. Develop specifics of the roofing system

A. 4, 2, 3 and 1
B. 2, 3, 4 and 1
C. 3, 4, 1 and 2
D. 1, 2, 4 and 3

QUESTION 55

Which of the following describes a "roof system"?

A. A
B. B
C. C
D. D

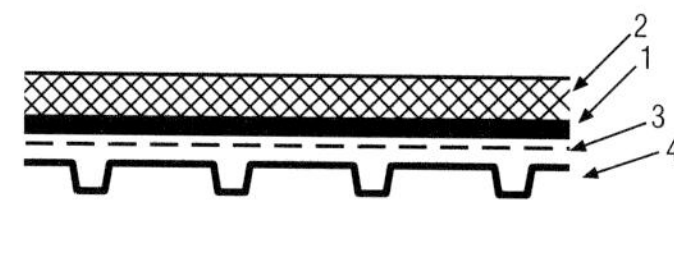

A

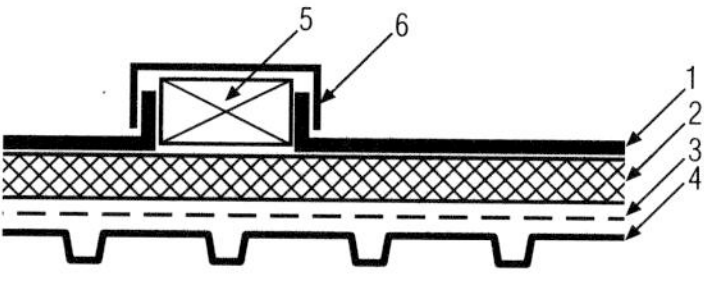

C

1
2
3

B

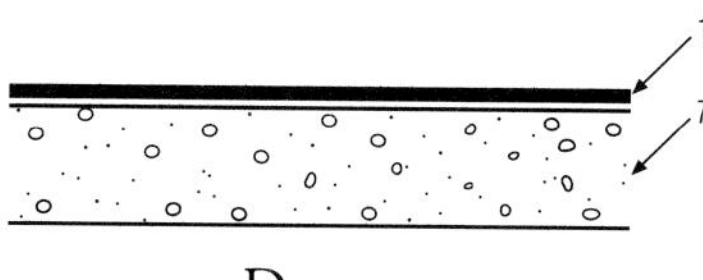

D

NOTES:

① **Membrane Roofing**
② **Insulation (if present)**
③ **Vapor Retarder (if present)**
④ **Metal Deck**
⑤ **Pressure-treated Wood Blocking**
⑥ **Cap Flashing**
⑦ **Concrete Deck**

QUESTION 56

Which of the following properties of bitumen are enhanced by the addition of polymers?

1. Color
2. Density
3. Elasticity
4. Flexibility

A. 1 and 3
B. 2 and 4
C. 3 and 4
D. 1 and 2

QUESTION 57

What three types of membrane roofings may have metal foil as surface treatment?

A. APP, SBS and SIS
B. SBS, SIS and SEBS
C. SIS, SEBS and APP
D. SEBS, APP and SBS

QUESTION 58

Which membrane roofing material is self-adhering?

A. SEBS
B. SBS
C. APP
D. SIS

QUESTION 59

Name two types of single-ply membranes.

1. Thermoset
2. Polyurethane
3. Thermoplastic
4. Modified bitumen

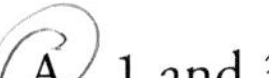

A. 1 and 3
B. 2 and 4
C. 3 and 4
D. 1 and 2

QUESTION 60

Which membrane would be best placed under a commercial kitchen roof exhaust fan?

A. EIP
B. NBP
C. TPO
D. CPA

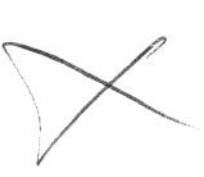

QUESTION 61

When an aggregate ballasted single-ply membrane is specified, what ballast load should typically be used for the ballast weight for the dead load calculations?

A. 10 psf, #1 aggregate
B. 13 psf, #2 aggregate
C. 13 psf, #4 aggregate
D. 17 psf, #6 aggregate

QUESTION 62

SPF systems are exceptionally thermally efficient, have good wind resistance and are not in imminent danger of leaking if the coating weathers away.

A. True
B. False

QUESTION 63

The design for reroofing is different from the design of a roof for a new building because of the

1. degree of slope of the roof deck.
2. possible deterioration of the deck.
3. effects of weather affecting the membrane.
4. possibility of friable asbestos below the deck.

A. 1 and 3
B. 1 and 4
C. 2 and 3
D. 2 and 4

QUESTION 64

The purpose of a nondestructive evaluation (NDE) is to determine the

A. dew point.
B. type of fasteners.
C. moisture content.
D. amount of ballast.

QUESTION 65

In a properly designed reroofing system, if insulation is added below the deck, calculations will show the dew point relocated to

A. above the deck.
B. below the deck.
C. above the insulation.
D. below the insulation.

QUESTION 66

Tear-off refers to

A. uplift.
B. edge failure.
C. wind blow-off.
D. removal of the roof.

QUESTION 67

Moisture in a roof can be detected with infrared thermography equipment which

A. creates an electrical field.
B. records the dew point at the deck.
C. detects surface temperature changes.
D. counts neutrons striking hydrogen atoms.

QUESTION 68

A minimum number of samples recommended for laboratory evaluation of asphalt on an existing deck is

A. two.
B. three.
C. four.
D. five.

QUESTION 69

Asphalt roof cement and coatings that contain asbestos are still permitted for new installations.

A. True
B. False

QUESTION 70

The intent of these details is to prevent

A. leaks at the roof edge.
B. roof overload (*ASCE 7*).
C. roof membrane blow-off.
D. uneven thermal expansion.

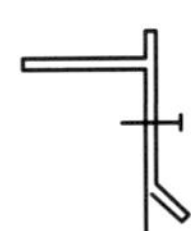 

QUESTION 71

Virtually every warranty issued by manufacturers covers only repair of leaks caused by defects of the materials.

A. True
B. False

QUESTION 72

Which membrane should the architect specify when a welded seam is desired for increased reliability?

A. PIB
B. SPF
C. PVC
D. EPDM

QUESTION 73

What roof insulation might be a very good choice for a sustainable roofing system?

A. Cellular glass
B. Fiber board
C. Polystyrene
D. Polyisocyanurate

LOW-SLOPE ROOFING II QUIZ

QUESTION 74

It is good practice to replace a roof with ponding water with a single-ply rubber membrane in order to save cost and avoid the need for positive slope to roof drains.

A. True
B. False

QUESTION 75

Liquid-applied membranes have exceptionally good resistance to

A. wind.
B. missiles.
C. puncture.
D. foot traffic.

QUESTION 76

Moisture entrapment within the roofing system can

1. detach the base sheet.
2. degrade the roof deck.
3. cause membrane creep.
4. enhance uplift resistance.

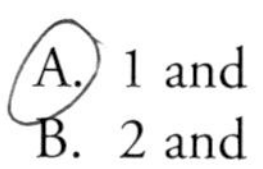

A. 1 and 2
B. 2 and 3
C. 3 and 4
D. 1 and 4

QUESTION 77

The following primary elements are among six included in the considerations for a sustainable roofing concept.

1. Design, installation
2. Materials, life-cycle cost
3. Service life, installed cost per unit of R
4. Thermal performance, material disposition

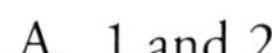

A. 1 and 2
B. 2 and 3
C. 3 and 4
D. 1 and 4

QUESTION 78

When the concept of "gradle" is used during the lifetime of a sustainable roof, the resulting reflectivity is typically high.

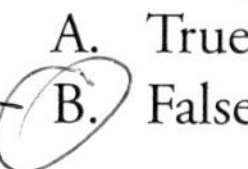

A. True
B. False

QUESTION 79

During the design of a sustainable ¼-inch-per-foot (2 percent) low-slope roof for an elementary school in rural southern Illinois, the architect considered several possible roof/insulation applications. Which system, based on your limited knowledge of the building and its site, would be his/her best choice?

A. Fiberglass insulation with metal panels
B. Extruded polystyrene insulation with a BUR
C. Polyisocyanurate insulation with an adhered EPDM membrane
D. Expanded polystyrene insulation with a ballasted EPDM membrane

QUESTION 80

Which of the following criteria can achieve thermal efficiency?

1. Highly reflective roof surface
2. Cap sheet layer on the surface
3. Roof assembly with a low U-value
4. Fill the parapet wall cavity with insulation

A. 1 and 3
B. 3 and 4
C. 1 and 2
D. 2 and 4

QUESTION 81

By using a "weighted evaluation methodology," it is not necessary for an architect to use subjectivity when making roofing system selection decisions.

A. True
B. False

QUESTION 82

An owner wants to build an addition to his/her office building (*see illustration*). Which roof system is the most suitable?

A. Cold-applied modified bitumen
B. EPDM ballasted
C. Metal panels
D. BUR

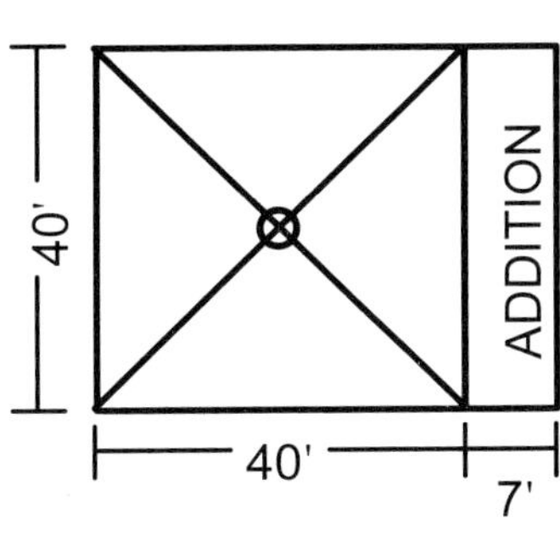

SECTION

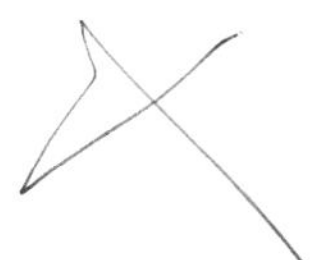